AF301263

Chocolate and Cocoa Recipes and Home Made Candy Recipes

Maria Parloa

Imprint

This book is part of TREDITION CLASSICS

Author: Maria Parloa
Cover design: Buchgut, Berlin – Germany

Publisher: tredition GmbH, Hamburg - Germany
ISBN: 978-3-8424-3482-0

www.tredition.com
www.tredition.de

Chocolate and Cocoa Recipes

By Miss Parloa

and

Home Made Candy Recipes

By Mrs. Janet McKenzie Hill

Compliments of Walter Baker & Co., Ltd.

ESTABLISHED DORCHESTER, MASS. 1780.

INDEX TO RECIPES

MISS PARLOA'S:

Plain Chocolate (For Drinking)

Chocolate, Vienna Style

Breakfast Cocoa

Chocolate Layer Cake

Chocolate Cake

Chocolate Marble Cake

Chocolate Glacé Cake

Chocolate Glacé

Chocolate Biscuit

Chocolate Wafers

Cinderella Cakes

Chocolate Éclairs

Chocolate Cookies

Chocolate Gingerbread

Vanilla Icing

Chocolate Icing

Chocolate Profiteroles

Chocolate Ice-cream

Chocolate Cream Pies

Chocolate Mousse

Chocolate Charlotte

Chocolate Bavarian Cream

Chocolate Cream

Chocolate Blanc-mange

Chocolate Cream Renversee

Baked Chocolate Custard

Chocolate Soufflé

MISS BURR'S:

Cracked Cocoa

For Three Gallons Breakfast Cocoa

Vanilla Chocolate with Whipped Cream

Chocolate Cream Pie

Chocolate Filling

Meringue

Cocoa Sticks

Cocoa Frosting

Cocoa Sauce

Cocoa Cake

Cocoa Meringue Pudding

Chocolate Almonds

Hot Chocolate Sauce

Cocoa Sponge Cake

Chocolate Frosting

Chocolate Cake; or, Devil's Food

Chocolate Ice-cream

Chocolate Whip

Cocoa Marble Cake

Chocolate Marble Cake

Chocolate Jelly

Cottage Pudding

Vanilla Sauce

Cocoanut Soufflé

Chocolate Sauce

Cocoa Biscuit

Cocoa Fudge

MISS ROBINSON'S:

Plain Chocolate 1 quart

Cocoa Sponge Cake

Cocoa Marble Cake

Cocoa Doughnuts

Cocoa Buns

MRS. RORER'S:

Chocolate Cake

MRS. LINCOLN'S:

Chocolate Caramels

MISS FARMER'S:

Chocolate Nougat Cake

Chocolate Cream Candy

MRS. ARMSTRONG'S:

Chocolate Pudding

Chocolate Charlotte

Chocolate Jelly with Crystallized Green Gages

MRS. BEDFORD'S:

Chocolate Crullers

Hot Cocoa Sauce for Ice-cream

Chocolate Macaroons

MRS. EWING'S:

Creamy Cocoa

Creamy Chocolate

MRS. HILL'S:

Cocoa Frappé

Chocolate Puffs

MRS. SALZBACHER'S:

Chocolate Hearts

Cocoa Charlotte

Chocolate Fudge with Fruit

Chocolate Macaroons

Petits Four

Potato Cake

Spanish Chocolate Cake

MRS. HILL'S CANDY RECIPES:

Peppermints, Chocolate Mints, etc.

Chocolate Caramel Walnuts

"Dot" Chocolate Coatings

Chocolate Dipped Peppermints

Ginger, Cherry, Apricot and Nut Chocolates

Chocolate Peanut Clusters

Chocolate Coated Almonds

Chocolate Dipped Parisian Sweets

Stuffed Dates, Chocolate Dipped

Chocolate Oysterettes

Turkish Paste with French Fruit

Chocolate Pecan Pralines

Vassar Fudge

Smith College Fudge

Wellesley Marshmallow Fudge

Double Fudge

Marbled Fudge

Fudge Hearts or Rounds

Marshmallow Fudge

Chocolate Dipped Fruit Fudge

Chocolate Cocoanut Cakes

Baker's Chocolate "Divinity"

Chocolate Nougatines

Plain Chocolate Caramels

Chocolate Nut Caramels

Ribbon Caramels

Fondant

Almond Chocolate Creams

Cherry Chocolate Creams

Chocolate Peppermints

Fig and Nut Chocolates

Chocolate Marshmallows

Maple Fondant Acorns

Chocolate Almond Bars

Almond Fondant Sticks

Almond Fondant Balls

Walnut Cream Chocolates

To Mold Candy for Dipping

Chocolate Butter Creams

Fondant for Soft Chocolate Creams

Rose Chocolate Creams

Pistachio Chocolate Creams

Surprise Chocolate Creams

Chocolate Peanut Brittle

Chocolate Pop Corn Balls

Chocolate Molasses Kisses

Cocoa and Chocolate

The term "Cocoa," a corruption of "Cacao," is almost universally used in English-speaking countries to designate the seeds of the small tropical tree known to botanists as THEOBROMA CACAO, from which a great variety of preparations under the name of cocoa and chocolate for eating and drinking are made. The name "Chocolatl" is nearly the same in most European languages, and is taken from the Mexican name of the drink, "Chocolate" or "Cacahuatl." The Spaniards found chocolate in common use among the Mexicans at the time of the invasion under Cortez in 1519, and it was introduced into Spain immediately after. The Mexicans not only used chocolate as a staple article of food, but they used the seeds of the cacao tree as a medium of exchange.

No better evidence could be offered of the great advance which has been made in recent years in the knowledge of dietetics than the remarkable increase in the consumption of cocoa and chocolate in this country. The amount retained for home consumption in 1860 was only 1,181,054 pounds—about 3-5 of an ounce for each inhabitant. The amount retained for home consumption for the year ending Dec. 31, 1908, was 93,956,721 pounds—over 16 ounces for each inhabitant.

Although there was a marked increase in the consumption of tea and coffee during the same period, the ratio of increase fell far below that of cocoa. It is evident that the coming American is going to be less of a tea and coffee drinker, and more of a cocoa and chocolate drinker. This is the natural result of a better knowledge of the laws of health, and of the food value of a beverage which nourishes the body while it also stimulates the brain.

Baron von Liebig, one of the best-known writers on dietetics, says:

"It is a perfect food, as wholesome as delicious, a beneficient restorer of exhausted power; but its quality must be good and it must be carefully prepared. It is highly nourishing and easily digested, and is fitted to repair wasted strength, preserve health, and prolong life. It agrees with dry temperaments and convalescents; with mothers who nurse their children; with those whose occupations oblige them to undergo severe mental strains; with public speakers, and with all those who give to work a portion of the time needed for sleep. It soothes both stomach and brain, and for this reason, as well as for others, it is the best friend of those engaged in literary pursuits."

M. Brillat-Savarin, in his entertaining and valuable work, *Physiologie du Goût*, says: "Chocolate came over the mountains [from Spain to France] with Anne of Austria, daughter of Philip III and queen of Louis XIII. The Spanish monks also spread the knowledge of it by the presents they made to their brothers in France. It is well known that Linnæus called the fruit of the cocoa tree *theobroma*, 'food for the gods.' The cause of this emphatic qualification has been sought, and attributed by some to the fact that he was extravagantly fond of chocolate; by others to his desire to please his confessor; and by others to his gallantry, a queen having first introduced it into France.

"The Spanish ladies of the New World, it is said, carried their love for chocolate to such a degree that, not content with partaking of it several times a day, they had it sometimes carried after them to church. This favoring of the senses often drew upon them the censures of the bishop; but the Reverend Father Escobar, whose metaphysics were as subtle as his morality was accommodating, declared, formally, that a fast was not broken by chocolate prepared with water; thus wire-drawing, in favor of his penitents, the ancient adage, '*Liquidum non frangit jejunium.*'

"Time and experience," he says further, "have shown that chocolate, carefully prepared, is an article of food as wholesome as it is agreeable; that it is nourishing, easy of digestion, and does not possess those qualities injurious to beauty with which coffee has been reproached; that it is excellently adapted to persons who are obliged to a great concentration of intellect; in the toils of the pulpit or the bar, and especially to travellers; that it suits the most feeble stomach; that excellent effects have been produced by it in chronic complaints, and that it is a last resource in affections of the pylorus.

"Some persons complain of being unable to digest chocolate; others, on the contrary, pretend that it has not sufficient nourishment, and that the effect disappears too soon. It is probable that the former have only themselves to blame, and that the chocolate which they use is of bad quality or badly made; for good and well-made chocolate must suit every stomach which retains the slightest digestive power.

"In regard to the others, the remedy is an easy one: they should reinforce their breakfast with a *pâté*, a cutlet, or a kidney, moisten the whole with a good draught of soconusco chocolate, and thank God for a stomach of such superior activity.

"This gives me an opportunity to make an observation whose accuracy may be depended upon.

"After a good, complete, and copious breakfast, if we take, in addition, a cup of well-made chocolate, digestion will be perfectly accomplished in three hours, and we may dine whenever we like. Out of zeal for science, and by dint of eloquence, I have induced many ladies to try this experiment. They all declared, in the beginning, that it would kill them; but they have all thriven on it and have not failed to glorify their teacher.

"The people who make constant use of chocolate are the ones who enjoy the most steady health, and are the least subject to a multitude of little ailments which destroy the comfort of life; their plumpness is also more equal. These are two advantages which every one may verify among his own friends, and wherever the practice is in use."

In corroboration of M. Brillat-Savarin's statement as to the value of chocolate as an aid to digestion, we may quote from one of Mme. de Sévigné's letters to her daughter:

"I took chocolate night before last to digest my dinner, in order to have a good supper. I took some yesterday for nourishment, so as to be able to fast until night. What I consider amusing about chocolate is that it acts according to the wishes of the one who takes it."

Chocolate appears to have been highly valued as a remedial agent by the leading physicians of that day. Christoph Ludwig Hoffman wrote a treatise entitled, "Potus Chocolate," in which he recommended it in many diseases, and instanced the case of Cardinal Richelieu, who, he stated, was cured of general atrophy by its use.

A French officer who served in the West Indies for a period of fifteen years, during the early part of the last century, wrote, as the result of his personal observations, a treatise on "The Natural History of Chocolate, Being a distinct and Particular Account of the Cacao Tree, its Growth and Culture, and the Preparation, Excellent Properties, and Medicinal Virtues of its Fruit," which received the approbation of the Regent of the Faculty of Medicine at Paris, and which was translated and published in London, in 1730. After describing the different methods of raising and curing the fruit and preparing it for food (which it is not worth while to reproduce here, as the methods have essentially changed since that time), he goes on to demonstrate, as the result of actual experiment, that chocolate is a substance "very temperate, very nourishing, and of easy digestion; very proper to repair the exhausted spirits and decayed strength; and very suitable to preserve the health and prolong the lives of old men...."

"I could produce several instances," he says, "in favor of this excellent nourishment; but I shall content myself with two only, equally certain and decisive, in proof of its goodness. The first is an experiment of chocolate's being taken for the only nourishment—made by a surgeon's wife of Martinico. She had lost, by a very deplorable accident, her lower jaw, which reduced her to such a condition that she did not know how to subsist. She was not capable of taking anything solid, and not rich enough to live upon jellies and nourishing broths. In this strait she determined to take three dishes of chocolate, prepared after the manner of the country, one in the morning, one at noon, and one at night. There chocolate is nothing else but cocoa kernels dissolved in hot water, with sugar, and seasoned with a bit of cinnamon. This new way of life succeeded so well that she has lived a long while since, more lively and robust than before this accident.

"I had the second relation from a gentleman of Martinico, and one of my friends not capable of a falsity. He assured me that in his neighborhood an infant of four months old unfortunately lost his nurse, and its parents not being able to put it to another, resolved, through necessity, to feed it with chocolate. The success was very happy, for the infant came on to a miracle, and was neither less healthy nor less vigorous than those who are brought up by the best nurses.

"Before chocolate was known in Europe, good old wine was called the milk of old men; but this title is now applied with greater reason to chocolate, since its use has become so common that it has been perceived that chocolate is, with respect to them, what milk is to infants. In reality, if one examines the nature of chocolate a little, with respect to the constitution of aged persons, it seems as though the one was made on purpose to remedy the defects of the other, and that it is truly the panacea of old age."

The three associated beverages, cocoa, tea, and coffee are known to the French as *aromatic* drinks. Each of these has its characteristic aroma. The fragrance and flavor are so marked that they cannot be imitated by any artificial products, although numerous attempts have been made in regard to all three. Hence the detection of adulteration is not a difficult matter. Designing persons, aware of the extreme difficulty of imitating these substances, have undertaken to employ lower grades, and, by manipulation, copy, as far as may be, the higher sorts. Every one knows how readily tea, and coffee, for that matter, will take up odors and flavors from substances placed near them. This is abundantly exemplified in the country grocery or general store, where the teas and coffees share in the pervasive fragrance of the cheese and kerosene. But perhaps it is not so widely understood that some of these very teas and coffees had been artificially flavored or corrected before they reached their destination in this country.

Cacao lends itself very readily to such preliminary treatment. In a first-class article, the beans should be of the highest excellence; they should be carefully grown on the plantation and there prepared with great skill, arriving in the factory in good condition. In the factory they should simply receive the mechanical treatment requisite to develop their high and attractive natural flavor and fragrance. They should be most carefully shelled after roasting and finely ground without concealed additions. This is the process in all honest manufactories of the cacao products.

Now, as a matter of fact, in the preparation of many of the cacao products on the market, a wholly different course has been pursued. Beans of poor quality are used, because of their cheapness, and in some instances they are only imperfectly, if at all, shelled before grinding. Chemical treatment is relied on to correct in part the odor and taste of such inferior goods, and artificial flavors, other than the time-honored natural vanilla and the like, are added freely. The detection of such imposition is easy enough to the expert, but is difficult to the novice; therefore the public is largely unable to discriminate between the good and the inferior, and it is perforce compelled to depend almost entirely on the character and reputation of the manufacturer.

A distinguished London Physician, in giving some hints concerning the proper preparation of cocoa, says:

"Start with a pure cocoa of undoubted quality and excellence of manufacture, and which bears the name of a respectable firm. This point is important, for there are many cocoas on the market which have been doctored by the addition of alkali, starch, malt, kola, hops, etc."

Baker's Breakfast Cocoa is absolutely pure, and, being ground to an extraordinary degree of fineness, is highly soluble. The analyst of the Massachusetts State Board of Health states in his recent valuable work on "Food Inspection and Analysis," that the treatment of cocoa with alkali for the purpose of producing a more perfect emulsion is objectionable, even if not considered as a form of adulteration. Cocoa thus treated is generally darker in color than the pure article. The legitimate means, he says, for making it as soluble as possible is to pulverize it very fine, so that particles remain in even suspension and form a smooth paste.

That is the way the Baker Cocoa is treated. It has received the Grand Prize—the highest award ever given in this country, and altogether 52 highest awards in Europe and America.

Choice Recipes
by
Miss Maria Parloa
SPECIALLY PREPARED FOR
WALTER BAKER & Co LTD.

PLAIN CHOCOLATE

For six people, use one quart of milk, two ounces of Walter Baker & Co.'s Premium No. 1 Chocolate, one tablespoonful of cornstarch, three tablespoonfuls of sugar, and two tablespoonfuls of hot water.

Mix the cornstarch with one gill of the milk. Put the remainder of the milk on to heat in the double-boiler. When the milk comes to the boiling point, stir in the cornstarch and cook for ten minutes. Have the chocolate cut in fine bits, and put it in a small iron or granite-ware pan; add the sugar and water, and place the pan over a hot fire. Stir constantly until the mixture is smooth and glossy. Add this to the hot milk, and beat the mixture with a whisk until it is frothy. Or, the chocolate may be poured back and forth from the boiler to a pitcher, holding high the vessel from which you pour. This will give a thick froth. Serve at once.

If you prefer not to have the chocolate thick, omit the cornstarch. If condensed milk is used, substitute water for the milk named above and add three tablespoonfuls of condensed milk when the chocolate is added.

CHOCOLATE, VIENNA STYLE

Use four ounces of Walter Baker & Co.'s Vanilla Chocolate, one quart of milk, three tablespoonfuls of hot water, and one tablespoonful of sugar.

Cut the chocolate in fine bits. Put the milk on the stove in the double-boiler, and when it has been heated to the boiling point, put the chocolate, sugar and water in a small iron or granite-ware pan, and stir over a hot fire until smooth and glossy. Stir this mixture into the hot milk, and beat well with a whisk. Serve at once, putting a tablespoonful of whipped cream in each cup and then filling up with the chocolate.

The plain chocolate may be used instead of the vanilla, but in that case use a teaspoonful of vanilla extract and three generous tablespoonfuls of sugar instead of one.

BREAKFAST COCOA

Walter Baker & Co.'s Breakfast Cocoa is powdered so fine that it can be dissolved by pouring boiling water on it. For this reason it is often prepared at the table. A small teaspoonful of the powder is put in the cup with a teaspoonful of sugar; on this is poured two-thirds of a cup of boiling water, and milk or cream is added to suit the individual taste. This is very convenient; but cocoa is not nearly so good when prepared in this manner as when it is boiled.

For six cupfuls of cocoa use two tablespoonfuls of the powder, two tablespoonfuls of sugar, half a pint of boiling water, and a pint and a half of milk. Put the milk on the stove in the double-boiler. Put the cocoa and sugar in a saucepan, and gradually pour the hot water upon them, stirring all the time. Place the saucepan on the fire and stir until the contents boil. Let this mixture boil for five minutes; then add the boiling milk and serve.

A gill of cream is a great addition to this cocoa.

Scalded milk may be used in place of boiled milk, if preferred. For flavoring, a few grains of salt and half a teaspoonful of vanilla extract may be added.

CHOCOLATE LAYER CAKE

Beat half a cupful of butter to a cream, and gradually beat into it one cupful of sugar. When this is light, beat in half a cupful of milk, a little at a time, and one teaspoonful of vanilla. Beat the whites of six eggs to a stiff froth. Mix half a teaspoonful of baking powder with two scant cupfuls of sifted flour. Stir the flour and whites of eggs alternately into the mixture. Have three deep tin plates well buttered, and spread two-thirds of the batter in two of them.

Into the remaining batter stir one ounce of Walter Baker & Co.'s Premium No. 1 Chocolate, melted, and spread this batter in the third plate. Bake the cakes in a moderate oven for about twenty minutes. Put a layer of white cake on a large plate, and spread with white icing. Put the dark cake on this, and also spread with white icing. On this put the third cake. Spread with chocolate icing.

TO MAKE THE ICING. Put into a granite-ware saucepan two gills of sugar and one of water, and boil gently until bubbles begin to come from the bottom—say, about five minutes. Take from the fire instantly. Do not stir or shake the sugar while it is cooking. Pour the hot syrup in a thin stream into the whites of two eggs that have been beaten to a stiff froth, beating the mixture all the time. Continue to beat until the icing is thick. Flavor with one teaspoonful of vanilla. Use two-thirds of this as a white icing, and to the remaining third add one ounce of melted chocolate. To melt the chocolate, shave it fine and put in a cup, which is then to be placed in a pan of boiling water.

CHOCOLATE CAKE

For two sheets of cake, use three ounces of Walter Baker & Co.'s Premium No. 1 Chocolate, three eggs, one cupful and three-fourths of sifted pastry flour, one cupful and three-fourths of sugar, half a cupful of butter, half a cupful of milk, half a teaspoonful of vanilla extract, one teaspoonful and a half of baking powder.

Grate the chocolate. Beat the butter to a cream, and gradually beat in the sugar. Beat in the milk and vanilla, then the eggs (already well beaten), next the chocolate, and finally the flour, in which the baking powder should be mixed. Pour into two well buttered shallow cake pans. Bake for twenty-five minutes in a moderate oven. Frost or not, as you like.

CHOCOLATE MARBLE CAKE

Put one ounce of Walter Baker & Co.'s Chocolate and one table-spoonful of butter in a cup, and set this in a pan of boiling water. Beat to a cream half a cupful of butter and one cupful of sugar. Gradually beat in half a cupful of milk. Now add the whites of six eggs beaten to a stiff froth, one teaspoonful of vanilla, and a cupful and a half of sifted flour, in which is mixed one teaspoonful of baking powder. Put about one-third of this mixture into another bowl, and stir the melted butter and chocolate into it. Drop the white-and-brown mixture in spoonfuls into a well buttered deep cake pan, and bake in a moderate oven for about forty-five minutes; or, the cake can be baked in a sheet and iced with a chocolate or white icing.

CHOCOLATE GLACÉ CAKE

Beat to a cream a generous half cupful of butter, and gradually beat into this one cupful of sugar. Add one ounce of Walter Baker & Co.'s Premium No. 1 Chocolate, melted; also two unbeaten eggs. Beat vigorously for five minutes; then stir in half a cupful of milk, and lastly, one cupful and a half of flour, with which has been mixed one generous teaspoonful of baking powder. Flavor with one teaspoonful of vanilla. Pour into a buttered, shallow cake pan, and bake for half an hour in a moderate oven. When cool, spread with glacé frosting.

GLACÉ FROSTING. Put half a cupful of sugar and three table-spoonfuls of water in a small saucepan. Stir over the fire until the sugar is nearly melted. Take the spoon from the pan before the sugar really begins to boil, because it would spoil the icing if the syrup were stirred after it begins to boil. After boiling gently for four minutes, add half a teaspoonful of vanilla extract, but do not stir; then set away to cool. When the syrup is about blood warm, beat it with a wooden spoon until thick and white. Now put the saucepan in another with boiling water, and stir until the icing is thin enough to pour. Spread quickly on the cake.

CHOCOLATE GLACÉ

After making a glacé frosting, dissolve one ounce of Walter Baker & Co.'s Premium No. 1 Chocolate in a cup, and put it with the frosting, adding also a tablespoonful of boiling water.

CHOCOLATE BISCUIT

Cover three large baking pans with paper that has been well oiled with washed butter. Over these dredge powdered sugar. Melt in a cup one ounce of Walter Baker & Co.'s Premium No. 1 Chocolate. Separate the whites and yolks of four eggs. Add to the yolks a generous half cupful of powdered sugar, and beat until light and firm. Add the melted chocolate, and beat a few minutes longer. Beat the whites of the eggs to a stiff, dry froth. Measure out three-fourths of a cupful of sifted flour, and stir it and the whites into the yolks. The whites and flour must be cut in as lightly as possible, and with very little stirring. Drop the mixture in teaspoonfuls on the buttered paper. Sprinkle powdered sugar over the cakes, and bake in a slow oven for about fourteen or fifteen minutes. The mixture can be shaped like lady fingers, if preferred.

CHOCOLATE WAFERS

Grate four ounces of Walter Baker & Co.'s Premium No. 1 Chocolate, and mix with it two tablespoonfuls of flour and one-fourth of a teaspoonful each of cinnamon, cloves and baking powder. Separate six eggs. Add one cupful of powdered sugar to the yolks, and beat until very light; then add the grated yellow rind and the juice of half a lemon, and beat five minutes longer. Now add the dry mixture, and with a spoon lightly cut in the whites, which are first to be beaten to a stiff froth. Pour the mixture into buttered shallow pans, having it about half an inch thick. Bake in a moderate oven for half an hour. When the cake is cool, spread a thin layer of currant jelly over one sheet, and place the other sheet on this. Ice with vanilla icing; and when this hardens, cut in squares. It is particularly nice to serve with ice-cream.

CINDERELLA CAKES

Use two eggs, one cupful of sugar, one cupful and a quarter of flour, one gill of cold water, one tablespoonful of lemon juice, one teaspoonful of baking powder, one ounce of Walter Baker & Co.'s Premium No. 1 Chocolate, half a tumbler of any kind of jelly, and chocolate icing the same as for éclairs.

Separate the eggs, and beat the yolks and sugar together until light. Beat the whites until light, and then beat them with yolks and sugar and grated chocolate. Next beat in the lemon juice and water, and finally the flour, in which the baking powder should be mixed. Beat for three minutes, and then pour the batter into two pans, and bake in a moderate oven for about eighteen minutes. When done, spread one sheet of cake with the jelly, and press the other sheet over it; and when cold, cut into little squares and triangular pieces. Stick a wooden toothpick into each of these pieces and dip each one into the hot icing, afterwards removing the toothpick, of course.

CHOCOLATE ÉCLAIRS

Into a granite-ware saucepan put half a pint of milk, two well-rounded tablespoonfuls of butter, and one tablespoonful of sugar, and place on the stove. When this boils up, add half a pint of sifted flour, and cook for two minutes, beating well with a wooden spoon. It will be smooth and velvety at the end of that time. Set away to cool; and when cool, beat in four eggs, one at a time. Beat vigorously for about fifteen minutes. Try a small bit of the paste in the oven; and if it rises in the form of a hollow ball, the paste is beaten enough; whereas, if it does not, beat a little longer. Have tin sheets or shallow pans slightly buttered. Have ready, also, a tapering tin tube, with the smaller opening about three-quarters of an inch in diameter. Place this in the small end of a conical cotton pastry bag. Put the mixture in the bag, and press out on buttered pans, having each éclair nearly three inches long. There should be eighteen, and they must be at least two inches apart, as they swell in cooking. Bake in a moderately hot oven for about twenty-five minutes. Take from the oven, and while they are still warm coat them with chocolate. When cold, cut open on the side, and fill with either of the following described preparations: —

FILLING NO. 1. — Mix in a bowl half a pint of rich cream, one teaspoonful of vanilla, and four tablespoonfuls of sugar. Place the bowl in a pan of ice-water, and beat the cream until light and firm, using either an egg-beater or a whisk.

FILLING NO. 2. — Put half a pint of milk into a double-boiler, and place on the fire. Beat together until very light one level tablespoonful of flour, half a cupful of sugar, and one egg. When the milk boils, stir in this mixture. Add one-eighth of a teaspoonful of salt, and cook for fifteen minutes, stirring often. When cold, flavor with one teaspoonful of vanilla.

ICING FOR ÉCLAIRS. — Put in a small granite-ware pan half a pint of sugar and five tablespoonfuls of cold water. Stir until the sugar is partially melted, and then place on the stove, stirring for half a minute. Take out the spoon, and watch the sugar closely. As soon as it boils, take instantly from the fire and pour upon a meat-platter. Let this stand for eight minutes. Meantime, shave into a cup one ounce of Walter Baker & Co.'s Premium No. 1 Chocolate, and put it on the fire in a pan of boiling water. At the end of eight minutes stir the sugar with a wooden spoon until it begins to grow white and to thicken. Add the melted chocolate quickly, and continue stirring until the mixture is thick. Put it in a small saucepan, and place on the fire in another pan of hot water. Stir until so soft that it will pour freely. Stick a skewer into the side of an éclair, and dip the top in the hot chocolate. Place on a plate, and continue until all the éclairs are "glacéd." They will dry quickly. Do not stir the sugar after the first half minute, and do not scrape the sugar from the saucepan into the platter. All the directions must be strictly followed.

CHOCOLATE COOKIES

Beat to a cream half a cupful of butter and one tablespoonful of lard. Gradually beat into this one cupful of sugar; then add one-fourth of a teaspoonful of salt, one teaspoonful of cinnamon, and two ounces of Walter Baker & Co.'s Premium No. 1 Chocolate, melted. Now add one well-beaten egg, and half a teaspoonful of soda dissolved in two tablespoonfuls of milk. Stir in about two cupfuls and a half of flour. Roll thin, and, cutting in round cakes, bake in a rather quick oven. The secret of making good cookies is the use of as little flour as will suffice.

CHOCOLATE GINGERBREAD

Mix in a large bowl one cupful of molasses, half a cupful of sour milk or cream, one teaspoonful of ginger, one of cinnamon, half a teaspoonful of salt. Dissolve one teaspoonful of soda in a teaspoonful of cold water; add this and two tablespoonfuls of melted butter to the mixture. Now stir in two cupfuls of sifted flour, and finally add two ounces of Walter Baker & Co.'s Chocolate and one tablespoonful of butter, melted together. Pour the mixture into three well-buttered, deep tin plates, and bake in a moderately hot oven for about twenty minutes.

VANILLA ICING

Break the white of one large egg into a bowl, and gradually beat into it one cupful of confectioners' sugar. Beat for three minutes, add half a teaspoonful of vanilla extract, and spread thinly on the cakes.

CHOCOLATE ICING

Make a vanilla icing, and add one tablespoonful of cold water to it. Scrape fine one ounce of Walter Baker & Co.'s Premium No. 1 Chocolate, and put it in a small iron or granite-ware saucepan, with two tablespoonfuls of confectioners' sugar and one tablespoonful of hot water. Stir over a hot fire until smooth and glossy, then add another tablespoonful of hot water. Stir the dissolved chocolate into the vanilla icing.

CHOCOLATE PROFITEROLES

Shave into a cup one ounce of Walter Baker & Co.'s Premium No. 1 Chocolate, and put the cup into a pan of boiling water. Make a paste the same as for éclairs, save that instead of one tablespoonful of sugar three must be used.

As soon as the paste is cooked, beat in the melted chocolate. When cold, add the eggs, and beat until light. Drop this batter on lightly buttered pans in round cakes, having about a dessertspoonful in each cake. Bake for about twenty minutes in a moderately hot oven. Serve either hot or cold, with whipped cream prepared the same as for Filling No. 1 for éclairs. Heap the cream in the center of a flat dish, and arrange the profiteroles around it.

CHOCOLATE ICE-CREAM

For about two quarts and a half of cream use a pint and a half of milk, a quart of thin cream, two cupfuls of sugar, two ounces of Walter Baker & Co.'s Premium No. 1 Chocolate, two eggs, and two heaping tablespoonfuls of flour.

Put the milk on to boil in a double-boiler. Put the flour and one cupful of the sugar in a bowl; add the eggs, and beat the mixture until light. Stir this into the boiling milk, and cook for twenty minutes, stirring often.

Scrape the chocolate, and put it in a small saucepan. Add four tablespoonfuls of sugar (which should be taken from the second cupful) and two tablespoonfuls of hot water. Stir over a hot fire until smooth and glossy. Add this to the cooking mixture.

When the preparation has cooked for twenty minutes, take it from the fire and add the remainder of the sugar and the cream, which should be gradually beaten into the hot mixture. Set away to cool, and when cold, freeze.

CHOCOLATE CREAM PIES

Beat to a cream half a cupful of butter and a cupful and a quarter of powdered sugar. Add two well-beaten eggs, two tablespoonfuls of wine, half a cupful of milk, and a cupful and a half of sifted flour, with which has been mixed a teaspoonful and a half of baking powder. Bake this in four well-buttered, deep, tin plates for about fifteen minutes in a moderate oven.

Put half a pint of milk in the double-boiler, and on the fire. Beat together the yolks of two eggs, three tablespoonfuls of powdered sugar, and a level tablespoonful of flour. Stir this mixture into the boiling milk, beating well. Add one-sixth of a teaspoonful of salt, and cook for fifteen minutes, stirring often. When cooked, flavor with half a teaspoonful of vanilla extract. Put two of the cakes on two large plates, spread the cream over them, and lay the other two cakes on top. Beat the whites of the two eggs to a stiff froth, and then beat into them one cupful of powdered sugar and one teaspoonful of vanilla. Shave one ounce of Walter Baker & Co.'s Premium No. 1 Chocolate, and put it in a small pan with two tablespoonfuls of sugar and one tablespoonful of boiling water. Stir over a hot fire until smooth and glossy. Now add three tablespoonfuls of cream or milk, and stir into the beaten egg and sugar. Spread on the pies and set away for a few hours.

CHOCOLATE MOUSSE

Put a three-quart mould in a wooden pail, first lining the bottom with fine ice and a thin layer of coarse salt. Pack the space between the mould and the pail solidly with fine ice and coarse salt, using two quarts of salt and ice enough to fill the space. Whip one quart of cream, and drain it in a sieve. Whip again all the cream that drains through. Put in a small pan one ounce of Walter Baker & Co.'s Premium No. 1 Chocolate, three tablespoonfuls of sugar and one of boiling water, and stir over a hot fire until smooth and glossy. Add three tablespoonfuls of cream. Sprinkle a cupful of powdered sugar over the whipped cream. Pour the chocolate in a thin stream into the cream, and stir gently until well mixed. Wipe out the chilled mould, and turn the cream into it. Cover, and then place a little ice lightly on top. Wet a piece of carpet in water, and cover the top of the pail. Set away for three or four hours; then take the mould from the ice, dip it in cold water, wipe, and then turn the mousse out on a flat dish.

CHOCOLATE CHARLOTTE

Soak a quarter of a package of gelatine in one-third of a cupful of cold water for two hours. Whip one pint of cream to a froth, and put it in a bowl, which should be placed in a pan of ice-water. Put half an ounce of shaved chocolate in a small pan with two tablespoonfuls of sugar and one of boiling water, and stir over the hot fire until smooth and glossy. Add to this a gill of hot milk and the soaked gelatine, and stir until the gelatine is dissolved. Sprinkle a generous half cupful of powdered sugar over the cream. Now add the chocolate and gelatine mixture, and stir gently until it begins to thicken. Line a quart charlotte-mould with lady fingers, and when the cream is so thick that it will just pour, turn it gently into the mould. Place the charlotte in a cold place for an hour or more, and, at serving time, turn out on a flat dish.

CHOCOLATE BAVARIAN CREAM

For one large mould of cream, use half a package of gelatine, one gill of milk, two quarts of whipped cream, one gill of sugar, and two and a half ounces of Walter Baker & Co.'s Chocolate.

Soak the gelatine in cold water for two hours. Whip and drain the cream, scrape the chocolate, and put the milk on to boil. Put the chocolate, two tablespoonfuls of sugar and one of hot water in a small saucepan, and stir on a hot fire until smooth and glossy. Stir this into the hot milk. Now add the soaked gelatine and the remainder of the sugar. Strain this mixture into a basin that will hold two quarts or more. Place the basin in a pan of ice-water, and stir until cold, when it will begin to thicken. Instantly begin to stir in the whipped cream, adding half the amount at first. When all the cream has been added, dip the mould in cold water and turn the cream into it. Place in the ice-chest for an hour or more.

At serving-time dip the mould in tepid water. See that the cream will come from the sides of the mould, and turn out on a flat dish. Serve with whipped cream.

CHOCOLATE CREAM

Soak a box of gelatine in half a pint of cold water for two hours. Put one quart of milk in the double-boiler, and place on the fire. Shave two ounces of Walter Baker & Co.'s Premium No. 1 Chocolate, and put it in a small pan with four tablespoonfuls of sugar and two of boiling water. Stir over a hot fire until smooth and glossy, and then stir into the hot milk. Beat the yolks of five eggs with half a cupful of sugar. Add to the gelatine, and stir the mixture into the hot milk. Cook three minutes longer, stirring all the while. On taking from the fire, add two teaspoonfuls of vanilla and half a saltspoonful of salt. Strain, and pour into moulds that have been rinsed in cold water. Set away to harden, and serve with sugar and cream.

CHOCOLATE BLANC-MANGE

Put one quart of milk in the double-boiler, and place on the fire. Sprinkle into it one level tablespoonful of sea-moss farina. Cover, and cook until the mixture looks white, stirring frequently. It will take about twenty minutes. While the milk and farina are cooking, shave two ounces of Walter Baker & Co.'s Premium No. 1 Chocolate, and put it into a small pan with four tablespoonfuls of sugar and two of boiling water. Stir over a hot fire until smooth and glossy, then stir into the cooked mixture. Add a saltspoonful of salt and a teaspoonful of vanilla. Strain, and turn into a mould that has been rinsed in cold water. Set the mould in a cold place, and do not disturb it until the blanc-mange is cold and firm. Serve with sugar and cream.

CHOCOLATE CREAM RENVERSEE

Use one quart of milk, seven eggs, half a pint of sugar, one ounce of Walter Baker & Co.'s Premium No. 1 Chocolate, half a teaspoonful of salt. Put the milk on the fire in the double-boiler. Shave the chocolate, and put it in a small pan with three tablespoonfuls of the sugar and one of boiling water. Stir over a hot fire until smooth and glossy; then stir into the hot milk, and take the milk from the fire to cool.

Put three tablespoonfuls of sugar into a charlotte-mould that will hold a little more than a quart, and place on the stove. When the sugar melts and begins to smoke, move the mould round and round, to coat it with the burnt sugar, then place on the table. Beat together the remainder of the sugar, the eggs, and the salt. Add the cold milk and chocolate to the mixture, and after straining into the charlotte-mould, place in a deep pan, with enough tepid water to come nearly to the top of the mould. Bake in a moderate oven until firm in the center. Test the cream by running a knife through the center. If firm and smooth, it is done. It will take forty or forty-five minutes to cook. When icy-cold, turn on a flat dish. Serve with whipped cream that has been flavored with sugar and vanilla.

BAKED CHOCOLATE CUSTARD

For five small custards use one pint of milk, two eggs, one ounce of Walter Baker & Co.'s Premium No. 1 Chocolate, one-fourth of a teaspoonful of salt, and a piece of stick cinnamon about an inch long.

Put the cinnamon and milk in the double-boiler, place on the fire and cook for ten minutes. Shave the chocolate, and put it in a small pan with three tablespoonfuls of sugar and one of boiling water. Stir this over a hot fire until smooth and glossy, and then stir it into the hot milk, after which take the liquid mixture from the fire and cool.

Beat together with a spoon the eggs, salt and two tablespoonfuls of the sugar. Add the cooled milk and strain. Pour the mixture into the cups, which place in a deep pan. Pour into the pan enough tepid water to come nearly to the top of the cups. Bake in a moderate oven until firm in the center. It will take about half an hour. Test by running a knife through the center. If the custard is milky, it is not done. Serve very cold.

CHOCOLATE SOUFFLÉ

Half a pint of milk, two ounces of Walter Baker & Co.'s Chocolate, three tablespoonfuls of sugar, one rounding tablespoonful of butter, two tablespoonfuls of flour, four eggs.

Put the milk in the double-boiler, and place on the fire. Beat the butter to a soft cream, and beat the flour into it. Gradually pour the hot milk on this, stirring all the time. Return to the fire and cook for six minutes. Put the shaved chocolate, sugar, and two tablespoonfuls of water in a small pan over a hot fire, and stir until smooth and glossy. Stir this into the mixture in the double-boiler. Take from the fire and add the yolks of the eggs, well beaten; then set away to cool. When cool add the whites of the eggs, beaten to a stiff froth. Pour the batter into a well-buttered earthen dish that will hold about a quart, and cook in a moderate oven for twenty-two minutes. Serve immediately with vanilla cream sauce.

CHOCOLATE PUDDING

Reserve one gill of milk from a quart, and put the remainder on the fire in a double-boiler. Mix three tablespoonfuls of cornstarch with the cold milk. Beat two eggs with half a cupful of powdered sugar and half a teaspoonful of salt. Add this to the cornstarch and milk, and stir into the boiling milk, beating well for a minute. Shave fine two ounces of Walter Baker & Co.'s Premium No. 1 Chocolate, and put it into a small pan with four tablespoonfuls of sugar and two of boiling water. Stir over a hot fire until smooth and glossy; then beat into the hot pudding. Cook the pudding in all ten minutes, counting from the time the eggs and cornstarch are added. Serve cold with powdered sugar and cream. This pudding can be poured while hot into little cups which have been rinsed in cold water. At serving time turn out on a flat dish, making a circle, and fill the center of the dish with whipped cream flavored with sugar and vanilla.

The eggs may be omitted, in which case use one more tablespoonful of cornstarch.

CHOCOLATE MERINGUE PUDDING

For a small pudding use one pint of milk, two tablespoonfuls and a half of cornstarch, one ounce of Walter Baker & Co.'s Chocolate, two eggs, five tablespoonfuls of powdered sugar, one-fourth of a teaspoonful of salt, and half a teaspoonful of vanilla extract.

Mix the cornstarch with one gill of the milk. Put the remainder of the milk on to boil in the double-boiler. Scrape the chocolate. When the milk boils, add the cornstarch, salt, and chocolate, and cook for ten minutes. Beat the yolks of the eggs with three tablespoonfuls of sugar. Pour the hot mixture on this, and beat well. Turn into a pudding-dish that will hold about a quart, and bake for twenty minutes in a moderate oven.

Beat the whites of the eggs to a stiff, dry froth, and gradually beat in the remaining two tablespoonfuls of sugar and the vanilla. Spread this on the pudding, and return to the oven. Cook for fifteen minutes longer, but with the oven-door open. Serve either cold or hot.

MILTON PUDDING

Use one pint of stale bread broken in crumbs, one quart of milk, two eggs, half a teaspoonful of salt, half a teaspoonful of ground cinnamon, three tablespoonfuls of sugar and two ounces of Walter Baker & Co.'s Chocolate, grated. Put the bread, milk, cinnamon, and chocolate in a bowl, and soak for two or three hours. Beat together the eggs, sugar, and salt. Mash the soaked bread with a spoon, and add the egg mixture to the bread and milk. Pour into a pudding-dish, and bake in a slow oven for about forty minutes. Serve with an egg sauce or a vanilla cream sauce.

EGG SAUCE. — Beat the whites of two eggs to a stiff, dry froth; and beat into this, a little at a time, one cupful of powdered sugar. When smooth and light, add one teaspoonful of vanilla and the yolks of two eggs. Beat the mixture a little longer; then stir in one cupful of whipped cream or three tablespoonfuls of milk. Serve at once.

VANILLA CREAM SAUCE. — Beat to a cream three tablespoonfuls of butter, and gradually beat into this two-thirds of a cupful of powdered sugar. When this is light and creamy, add a teaspoonful of vanilla; then gradually beat in two cupfuls of whipped cream. Place the bowl in a pan of boiling water, and stir constantly for three minutes. Pour the sauce into a warm bowl, and serve.

SNOW PUDDING

Put a pint of milk in the double-boiler and on the fire. Mix three tablespoonfuls of cornstarch with a gill of milk and one-third of a teaspoonful of salt. Stir this into the milk when it boils. Beat the whites of four eggs to a stiff froth, and then gradually beat into them half a cupful of powdered sugar and one teaspoonful of vanilla. Add this to the cooking mixture, and beat vigorously for one minute. Rinse a mould in cold water, and pouring the pudding into it, set away to cool. At serving-time turn out on a flat dish, and serve with chocolate sauce.

CHOCOLATE SAUCE

Put one pint of milk in the double-boiler, and on the fire. Shave two ounces of Walter Baker and Co.'s Chocolate, and put it in a small pan with four tablespoonfuls of sugar and two of boiling water. Stir over the fire until smooth and glossy, and add to the hot milk. Beat together for eight minutes the yolks of four eggs, three tablespoonfuls of sugar, and a saltspoonful of salt, and then add one gill of cold milk.

Pour the boiling milk on this, stirring well. Return to the double-boiler, and cook for five minutes, stirring all the time. Pour into a cold bowl and set the bowl in cold water. Stir for a few minutes, and then occasionally until the sauce is cold.

This sauce is nice for cold or hot cornstarch pudding, bread pudding, cold cabinet pudding, snow pudding, etc. It will also answer for a dessert. Fill custard glasses with it, and serve the same as soft custard; or have the glasses two-thirds full, and heap up with whipped cream.

CHOCOLATE CANDY

One cupful of molasses, two cupfuls of sugar, one cupful of milk, one-half pound of chocolate, a piece of butter half the size of an egg. Boil the milk and molasses together, scrape the chocolate fine, and mix with just enough of the boiling milk and molasses to moisten; rub it perfectly smooth, then, with the sugar, stir into the boiling liquid; add the butter, and boil twenty minutes. Try as molasses candy, and if it hardens, pour into a buttered dish. Cut the same as nut candy.

CREAM CHOCOLATE CARAMELS

Mix together in a granite-ware saucepan half a pint of sugar, half a pint of molasses, half a pint of thick cream, one generous tablespoonful of butter, and four ounces of Walter Baker & Co.'s Premium No. 1 Chocolate. Place on the fire and stir until the mixture boils. Cook until a few drops of it will harden if dropped into ice-water; then pour into well-buttered pans, having the mixture about three-fourths of an inch deep. When nearly cold, mark into squares. It will take almost an hour to boil this in a granite-ware pan, but not half so long if cooked in an iron frying-pan. Stir frequently while boiling. The caramels must be put in a very cold place to harden.

SUGAR CHOCOLATE CARAMELS

Mix two cupfuls of sugar, three-fourths of a cupful of milk or cream, one generous tablespoonful of butter, and three ounces of Walter Baker & Co.'s Premium No. 1 Chocolate. Place on the fire and cook, stirring often, until a little of the mixture, when dropped in ice-water, will harden; then stir in one-fourth of a cupful of sugar and one tablespoonful of vanilla, and pour into a well-buttered pan, having the mixture about three-fourths of an inch deep. When nearly cold, mark it off in squares, and put in a cold place to harden. These caramels are sugary and brittle, and can be made in the hottest weather without trouble. If a deep granite-ware saucepan be used for the boiling, it will take nearly an hour to cook the mixture; but if with an iron frying-pan, twenty or thirty minutes will suffice.

CHOCOLATE CREAMS, No. 1

Beat the whites of two eggs to a stiff froth. Gradually beat into this two cupfuls of confectioners' sugar. If the eggs be large, it may take a little more sugar. Flavor with half a teaspoonful of vanilla, and work well. Now roll into little balls, and drop on a slightly buttered platter. Let the balls stand for an hour or more. Shave five ounces of Walter Baker & Co.'s Premium No. 1 Chocolate and put into a small bowl, which place on the fire in a saucepan containing boiling water. When the chocolate is melted, take the saucepan to the table, and drop the creams into the chocolate one at a time, taking them out with a fork and dropping them gently on the buttered dish. It will take half an hour or more to harden the chocolate.

CHOCOLATE CREAMS, No. 2

For these creams you should make a fondant in this way: put into a granite-ware saucepan one cupful of water and two of granulated sugar — or a pound of loaf sugar. Stir until the sugar is nearly melted, then place on the fire and heat slowly, but do not stir the mixture. Watch carefully and note when it begins to boil. When the sugar has been boiling for ten minutes, take up a little of it and drop in ice-water. If it hardens enough to form a soft ball when rolled between the thumb and finger, it is cooked enough. Take the saucepan from the fire instantly, and set in a cool, dry place. When the syrup is so cool that the finger can be held in it comfortably, pour it into a bowl, and stir with a wooden spoon until it becomes thick and white. When it begins to look dry, and a little hard, take out the spoon, and work with the hand until the cream is soft and smooth. Flavor with a few drops of vanilla, and, after shaping, cover with chocolate, as directed in the preceding recipe.

Caution. — Do not stir the syrup while it is cooking, and be careful not to jar or shake the saucepan.

CHOCOLATE CONES

Boil the sugar as directed for fondant in the recipe for Chocolate Creams, No. 2, but not quite so long—say about eleven minutes. The syrup, when tested, should be too soft to ball. When cold, pour into a bowl, and beat until thick and creamy. If properly boiled, it will not become thick enough to work with the hands.

Have six ounces of Walter Baker & Co.'s Premium No. 1 Chocolate melted in a bowl. Pour half of the creamed sugar into another bowl, and, after flavoring with a few drops of vanilla, add to it about one-third of the dissolved chocolate. Stir until thick and rather dry; then make into small cones, and drop on a slightly buttered platter. Put half of the remaining creamed sugar in a cup, and set in a saucepan containing boiling water. Flavor with vanilla, and stir over the fire until melted so much that it will pour from the spoon. Take the saucepan to the table and dip one-half the cones in, one at a time, just as the Chocolate Creams, No. 1, were dipped in the melted chocolate. If liked, a second coating may be given the cones. Now put the remainder of the creamed sugar on to melt, and add two tablespoonfuls of hot water to it. Stir the remainder of the melted chocolate into this, and if too thick to dip the candy in, add hot water, a few drops at a time, until the mixture is of the right consistency; then dip the rest of the cones in this.

GENESEE BON-BONS

Make the cream chocolate caramels, and get them quite firm by placing the pan on ice. Make the chocolate coating as directed for chocolate cones. Dip the caramels in this and put on a buttered dish.

CHOCOLATE SYRUP

Into a granite-ware saucepan put one ounce—three tablespoonfuls—of Walter Baker & Co.'s Soluble Chocolate, and gradually pour on it half a pint of boiling water, stirring all the time. Place on the fire, and stir until all the chocolate is dissolved. Now add one pint of granulated sugar, and stir until it begins to boil. Cook for three minutes longer, then strain and cool. When cool, add one tablespoonful of vanilla extract. Bottle, and keep in a cold place.

REFRESHING DRINKS FOR SUMMER

Put into a tumbler about two tablespoonfuls of broken ice, two tablespoonfuls of chocolate syrup, three tablespoonfuls of whipped cream, one gill of milk, and half a gill of soda-water from a syphon bottle, or Apollinaris water. Stir well before drinking. A tablespoonful of vanilla ice-cream is a desirable addition. It is a delicious drink, even if the soda or Apollinaris water and ice-cream be omitted. A plainer drink is made by combining the syrup, a gill and a half of milk, and the ice, shaking well.

A FEW SUGGESTIONS IN REGARD TO CHOCO-LATE

The best flavor to add to chocolate is vanilla; next to that, cinnamon. Beyond these two things one should use great caution, as it is very easy to spoil the fine natural flavor of the bean. Chocolate absorbs odors readily; therefore it should be kept in a pure, sweet atmosphere. As about eleven per cent. of the chocolate bean is starch, chocolate and cocoa are of a much finer flavor if boiled for a few minutes. Long boiling, however, ruins their flavor and texture.

(All measurements should be level.)

FORMULA FOR MAKING THREE GALLONS OF BREAKFAST COCOA

- ½ a pound of Walter Baker & Co.'s Cocoa,
- 1 ½ gallons of water, hot,
- 1 ½ gallons of milk, hot.

This should not be allowed to boil. Either make it in a large double-boiler, or a large saucepan or kettle over water. Mix the cocoa with enough cold water to make a paste, and be sure it is free from lumps. Heat together the milk and water, and pour in the cocoa; then cook at least an hour, stirring occasionally.

CRACKED COCOA

To one-third a cup of Baker's Cracked Cocoa (sometimes called "Cocoa Nibs") use three cups of cold water; cook slowly at least one hour—the longer the better. Then strain the liquid and add one cup (or more if desired) of milk, and serve very hot. Do not allow the mixture to boil after milk has been added.

VANILLA CHOCOLATE WITH WHIPPED CREAM

- One cake (½ a pound) of Walter Baker & Co.'s Vanilla Sweet Chocolate,
- 4 cups of boiling water,
- Pinch of salt,
- 4 cups of hot milk.

This must be made in a double-boiler. Put the chocolate, boiling water and salt in upper part of the double-boiler. Stir and beat with a wooden spoon until the chocolate is dissolved and smooth. Add the milk and when thoroughly hot, strain, and serve with unsweetened whipped cream. More cooking will improve it.

CHOCOLATE CREAM PIE

Line a pie plate with rich pie crust, putting on an extra edge of crust the same as for custard pie. Fill with the chocolate filling made after the following recipe. Bake in a hot oven until crust is done; remove, and when cool, cover with a meringue and brown very slowly in moderate oven.

CHOCOLATE FILLING

- 1 cup of milk,
- Pinch of salt,
- 1 ½ squares of Baker's Chocolate,
- 2 level tablespoonfuls of flour,
- 2 eggs (yolks),
- 5 tablespoonfuls of sugar (level),
- 1 teaspoonful of vanilla.

Put milk, salt and chocolate in upper part of the double-boiler, and when hot and smooth, stir in the flour, which has been mixed with enough cold milk to be thin enough to pour into the hot milk. Cook, stirring constantly, until it thickens; then let it cook eight or ten minutes. Mix the eggs and sugar together and pour the hot mixture over them, stirring well; put back in double-boiler and cook, stirring constantly one minute. Remove, and when cool add one teaspoonful of vanilla.

MERINGUE

- 2 eggs (whites),
- Pinch of salt,
- 4 level tablespoonfuls of sugar,
- 1 teaspoonful of vanilla.

Add salt to eggs and beat in a large shallow dish with fork or egg-whip until stiff and flaky and dish can be turned upside down. Beat in the sugar slowly, then the vanilla, and beat until the dish can be turned upside down.

COCOA STICKS

- 6 tablespoonfuls of butter,
- ¾ cup of sugar (scant),
- 1 egg,
- 1 tablespoonful of milk,
- 1 teaspoonful of vanilla or pinch of cinnamon,
- 5 teaspoonfuls of Baker's Cocoa,
- 1/8 teaspoonful of baking powder,
- 1 ¼ to 1 ½ cups of sifted pastry flour.

Cream the butter until soft; add the sugar gradually and beat well; add the beaten egg, milk and vanilla; mix thoroughly. Sift cocoa, baking powder, and a pinch of salt with about one-half cup of the flour; stir this into the mixture first, then use the remainder of the flour, and more if necessary, to make a firm dough that will not stick to the fingers. Set on the ice to harden. Sprinkle the board with cocoa and a very little sugar. Use small pieces of the dough at a time, toss it over the board to prevent sticking, roll out thin, cut in strips about one-half inch wide and three inches long. Place closely in pan and bake in moderately hot oven three or four minutes. Great care should be taken in the baking to prevent burning.

It is advisable to gather the scraps after each rolling, if soft, and set away to harden, for fear of getting in too much cocoa, thus making them bitter.

The colder and harder the dough is, the better it can be handled; therefore it can be made the day before using.

COCOA FROSTING

- 4 teaspoonfuls of Baker's Cocoa,
- 2 tablespoonfuls of cold water,
- 3 tablespoonfuls of hot water,
- ½ a teaspoonful of vanilla,
- About 1 ¾ cups of confectioners' sugar.

Put the cocoa in a small saucepan; add the cold water and stir until perfectly smooth; then the hot water, and cook for one or two minutes, add vanilla and a speck of salt, then stir in enough sugar to make it stiff enough to spread nicely. Beat until smooth and glossy and free from lumps.

If too thick, add a little cold water. If not thick enough, add a little sugar. Never make a frosting so stiff that it will have to be made smooth with a wet knife. It is better to let it run to the sides of the cake. For frosting sides of the cake, make a little stiffer.

This frosting never cracks as an egg frosting, but is hard enough to cut nicely.

COCOA SAUCE

- 2 tablespoonfuls of butter,
- 1 cup of boiling water,
- 2 tablespoonfuls of flour,
- 4 tablespoonfuls of sugar,
- 4 teaspoonfuls of Baker's Cocoa,
- 1 teaspoonful of vanilla.

Melt the butter in the saucepan; mix the flour and cocoa together and stir into the butter; add gradually the hot water, stirring and beating each time; cook until it thickens. Just before serving, add the sugar, vanilla and a pinch of salt, if necessary.

Use more cocoa if liked stronger. This sauce will be found excellent for cottage puddings, Dutch apple cakes, steamed apple puddings, etc.

COCOA CAKE

- ½ a cup of butter,
- ¾ a cup of milk,
- 1 cup of sugar,
- 6 level tablespoonfuls of Baker's Cocoa,
- 3 eggs,
- 2 level teaspoonfuls of baking powder,
- 1 teaspoonful of vanilla,
- 1 ½ or 2 cups of sifted pastry flour.

Cream the butter, stir in the sugar gradually, add the unbeaten eggs, and beat all together until very creamy. Sift together one-half cup of the flour, the cocoa and baking powder; use this flour first, then alternate the milk and remaining flour, using enough to make mixture stiff enough to drop from the spoon; add vanilla and beat until very smooth; then bake in loaf in moderately hot oven thirty-five or forty minutes.

Tests for baking cake. It is baked enough when:

1. It shrinks from the pan.
2. Touching it on the top, springs back.
3. No singing sound.

COCOA MERINGUE PUDDING

- 1 cup of milk,
- 2 eggs (yolks),
- 2 tablespoonfuls of flour,
- Pinch of salt,
- 4 teaspoonfuls of Baker's Cocoa,
- 3 tablespoonfuls of sugar,
- ½ a teaspoonful of vanilla.

Put the milk in the upper part of the double-boiler, and heat. Mix flour and cocoa together and soften in a little cold milk; mix until free from lumps. When the milk is hot, add the flour, and cook, stirring often, eight or ten minutes. Beat yolks of eggs lightly; add sugar and salt, and mix well. When mixture in double-boiler has cooked sufficiently, strain it over the mixture in the bowl. Put back in double-boiler and allow it to cook one or two minutes (stirring constantly), just enough to slightly thicken the eggs. Remove from the stove, and when cool add vanilla and put in the serving-dish. Cover with a meringue. Place dish on a board, put in the oven with the door open, and allow it to remain there for ten or fifteen minutes, and when the meringue will not stick to the fingers, close the door and let it brown slightly. This pudding can be eaten warm or cold, but is much better cold. This will serve four persons generously.

CHOCOLATE ALMONDS

Blanch the almonds by pouring boiling water on them, and let them stand two or three minutes. Roast them in oven. Dip them in the following recipe for chocolate coating, and drop on paraffine paper.

- ½ pound cake of Walter Baker's Vanilla Sweet Chocolate,
- 2 level tablespoonfuls of butter,
- 2 tablespoonfuls of boiling water.

Put chocolate in small saucepan over boiling water and when melted stir in butter and water. Mix well. If found to be too thick, add more water; if too thin, more chocolate.

HOT CHOCOLATE SAUCE

- 1 cup of boiling water,
- Pinch of salt,
- 1 square of chocolate,
- ½ a cup of sugar.

Cook all together slowly until it is the consistency of maple syrup, or thicker if desired. Just before serving, add one teaspoonful of vanilla. This will keep indefinitely, and can be reheated.

COCOA SPONGE CAKE

- 4 eggs,
- ¼ a cup of sugar,
- Pinch of salt,
- 4 tablespoonfuls of Baker's Cocoa,
- ½ a cup of sifted pastry flour,
- 1 teaspoonful of vanilla.

Separate yolks from whites of eggs; beat yolks in a small bowl with the Dover egg-beater until very thick; add sugar, salt and vanilla, and beat again until very thick. Sift cocoa and the flour together and stir very lightly into the mixture; fold in the stiffly beaten whites of the eggs, and bake in a loaf in a moderate oven until done.

Do not butter the pan, but when cake is baked, invert the pan; and when cool, remove the cake.

CHOCOLATE FROSTING

- 1 square of Baker's Chocolate,
- Pinch of salt,
- 5 tablespoonfuls of boiling water,
- 1 teaspoonful of vanilla,
- About three cups of sifted confectioners' sugar.

Melt chocolate in bowl over tea-kettle, add water, salt and vanilla, and when smooth add the sugar, and heat until very glossy. Make the frosting stiff enough to spread without using a wet knife. It will keep indefinitely.

CHOCOLATE CAKE, OR DEVIL'S FOOD

- 5 level tablespoonfuls of butter,
- 1 ¼ cups of sugar,
- 3 ½ squares of Baker's Chocolate, (melted),
- 3 eggs,
- 1 teaspoonful of vanilla,
- ¾ a cup of milk,
- 3 ½ level teaspoonfuls of baking powder,
- 1 ½ cups of sifted pastry flour.

Cream the butter, add sugar and chocolate, then the unbeaten eggs and vanilla, and beat together until very smooth. Sift the baking powder with one-half a cup of the flour, and use first; then alternate the milk and the remaining flour, and make the mixture stiff enough to drop from the spoon. Beat until very smooth and bake in loaf in moderate oven. For tests see Cocoa Cake recipe on page 25.

CHOCOLATE ICE-CREAM

- 1 quart of milk,
- Pinch of salt,
- 3 squares of Baker's Chocolate,
- 3 level tablespoonfuls of flour,
- 1 can of sweetened condensed milk,
- 3 eggs,
- 6 level tablespoonfuls of sugar,
- 3 teaspoonfuls of vanilla.

Put milk, salt and chocolate in double-boiler, and when milk is hot and chocolate has melted, stir in the flour, previously mixed in a little cold milk. Cook ten minutes, then pour this over the condensed milk, eggs and sugar mixed together; cook again for four minutes, stirring. Strain, and when cool add vanilla, and freeze.

CHOCOLATE WHIP

- 1 cup of milk,
- 1 ½ squares of Baker's Chocolate,
- Pinch of salt,
- 2 level tablespoonfuls of cornstarch,
- 2 eggs (yolks),
- 6 level tablespoonfuls of sugar,
- 2 teaspoonfuls of vanilla,
- 5 eggs (whites).

Put milk, chocolate and salt in double-boiler; mix cornstarch in a small quantity of cold milk, and stir into the hot milk when the chocolate has been melted; stir until smooth, then cook twelve minutes. Mix together the yolks of the eggs and sugar, then pour the hot mixture over it; cook again one or two minutes, stirring. When very cold, just before serving, add the vanilla and fold in the stiffly beaten whites of the eggs. Pile lightly in a glass dish and serve with lady fingers. A meringue can be made of the whites of the eggs and sugar, then folded in the chocolate mixture, but it does not stand as long.

COCOA MARBLE CAKE

- 6 level tablespoonfuls of butter,
- 1 cup of granulated sugar,
- 3 eggs,
- 1 teaspoonful of vanilla,
- ¾ a cup of milk.

Three level teaspoonfuls of baking powder, about one and three-quarter cups of sifted flour, or flour enough to make mixture stiff enough to drop from the spoon. Mix in the order given. Reserve one-third of this mixture and add to it four level tablespoonfuls of Baker's Cocoa and to the other one cup of shredded cocoanut. Bake thirty-five or forty minutes according to size and shape of pan.

CHOCOLATE MARBLE CAKE

This is the same as the Cocoa Marble Cake. Add to one-third of the mixture one and one-half squares of Baker's Chocolate in place of the cocoa, and one cup of chopped walnuts to the other part in place of the shredded cocoanut.

CHOCOLATE JELLY

- 1 pint of boiling water,
- 1/3 a package of gelatine,
- 2 pinches of salt,
- 2 level tablespoonfuls of sugar,
- 1 ½ squares of Baker's Chocolate,
- 1 teaspoonful of vanilla.

Put the water, salt and chocolate in a saucepan. Cook, stirring until the chocolate melts, then let it boil for three or five minutes. Soften the gelatine in a little cold water and pour the boiling mixture over it. Stir until dissolved, then add sugar and vanilla. Pour into a mould and set aside to harden, serve with cream and powdered sugar or sweetened whipped cream.

COTTAGE PUDDING

- 4 level tablespoonfuls of butter,
- 2 eggs,
- 1 cup of sugar,
- ¾ a cup of milk.

Two level teaspoonfuls of baking powder, one and three-quarter cups of sifted flour or enough to make mixture stiff enough to drop from the spoon. Bake in buttered gem pans in moderately hot oven twenty-three or twenty-five minutes. If the cake springs back after pressing a finger on the top, it shows that it is baked enough. To make a cocoa cottage pudding add to the above rule six level table-spoonfuls of cocoa. Serve with a vanilla sauce.

VANILLA SAUCE

- 2 level tablespoonfuls of butter,
- 1 cup of boiling water,
- 2 level tablespoonfuls of flour,
- 4 level tablespoonfuls of sugar,
- Pinch of salt,
- 1 teaspoonful of vanilla.

Melt butter in saucepan, add flour and salt and mix until smooth; add slowly the boiling water, stirring and beating well. Add sugar and milk.

COCOANUT SOUFFLÉ

- 1 cup of milk,
- 1 pinch of salt,
- 3 level tablespoonfuls of flour, softened in a little cold milk.
- 2 level tablespoonfuls of butter,
- 4 level tablespoonfuls of sugar,
- Yolks of 4 eggs,
- 1 teaspoonful of vanilla,
- 1 cup of shredded cocoanut,
- Whites of 4 eggs.

Heat milk, add salt and flour and cook ten minutes after it has thickened. Mix together, butter, sugar and yolks of eggs. Pour hot mixture over, stirring well and set aside to cool. Add vanilla and cocoanut. Lastly fold in the stiffly beaten whites of the eggs. Bake in buttered pan, in moderate oven until firm. Serve hot with Chocolate Sauce.

CHOCOLATE SAUCE

- 2 level tablespoonfuls of butter,
- 1 level tablespoonful of flour,
- Pinch of salt,
- 1 cup of boiling water,
- 1 square of Baker's Chocolate,
- 4 level tablespoonfuls of sugar,
- 1 teaspoonful of vanilla.

Melt butter in saucepan, add dry flour and salt and mix until smooth, then add slowly the hot water, beating well. Add the square of chocolate and sugar and stir until melted. Add vanilla, just before serving.

COCOA BISCUIT

- 2 cups or 1 pint of sifted flour,
- 3 level teaspoonfuls of baking powder,
- ½ a teaspoonful of salt,
- 2 level tablespoonfuls of sugar,
- 4 level tablespoonfuls of Baker's Cocoa,
- 2 level tablespoonfuls of butter or lard,
- 2/3 a cup of milk or enough to make a firm but not a stiff dough.

Sift all the dry ingredients together, rub in the butter with the tips of the fingers. Stir in the required amount of milk. Turn out on slightly floured board, roll or pat out the desired thickness, place close together in pan and bake in very hot oven ten or fifteen minutes.

COCOA FUDGE

- ½ a cup of milk,
- 3 level tablespoonfuls of butter,
- 2 ½ cups of powdered sugar,
- 6 tablespoonfuls of Baker's Cocoa,
- Pinch of salt,
- 1 teaspoonful of vanilla.

Mix all ingredients together but vanilla; cook, stirring constantly, until it begins to boil, then cook slowly, stirring occasionally, eight or ten minutes, or until it makes a firm ball when dropped in cold water. When cooked enough, add the vanilla and beat until it seems like very cold molasses in winter. Pour into a buttered pan; when firm, cut in squares. Great care must be taken not to beat too much, because it cannot be poured into the pan, and will not have a gloss on top.

Miss M.E. Robinson's Recipes

PLAIN CHOCOLATE

- 1 ounce or square of Baker's Premium Chocolate,
- 3 tablespoonfuls of sugar,
- 1/8 a teaspoonful of salt,
- 1 pint of boiling water,
- 1 pint of milk.

Place the chocolate, sugar and salt in the agate chocolate-pot or saucepan, add the boiling water and boil three minutes, stirring once or twice, as the chocolate is not grated. Add the milk and allow it time to heat, being careful not to boil the milk, and keep it closely covered, as this prevents the scum from forming. When ready to serve turn in chocolate-pitcher and beat with Dover egg-beater until light and foamy.

COCOA DOUGHNUTS

One egg, one-half a cup of sugar, one-half a cup of milk, one-quarter teaspoonful of salt, one-quarter teaspoonful of cinnamon extract (Burnett's), two cups of flour, one-quarter cup of Baker's Breakfast Cocoa, two teaspoonfuls of baking powder. Mix in the order given, sifting the baking powder and cocoa with the flour. Roll to one-third an inch in thickness, cut and fry.

COCOA SPONGE CAKE

- 3 eggs,
- 1 ½ cups of sugar,
- ½ a cup of cold water,
- 1 teaspoonful of vanilla,
- 1 ¾ cups of flour,
- ¼ a cup of Baker's Cocoa,
- 2 teaspoonfuls of baking powder,
- 1 teaspoonful of cinnamon.

Beat yolks of eggs light, add water, vanilla and sugar; beat again thoroughly; then add the flour, with which the baking powder, cocoa and cinnamon have been sifted. Fold in the stiffly beaten whites of the eggs. Bake in a rather quick oven for twenty-five or thirty minutes.

COCOA MARBLE CAKE

- 1/3 a cup of butter,
- 1 cup of sugar,
- 1 egg,
- ½ a cup of milk,
- 1 teaspoonful of vanilla,
- 2 cups of flour,
- 2 teaspoonfuls of baking powder,
- 3 tablespoonfuls of Baker's Cocoa.

Cream the butter, add sugar, vanilla and egg; beat thoroughly, then add flour (in which is mixed the baking powder) and milk, alternately, until all added. To one-third of the mixture add the cocoa, and drop the white and brown mixture in spoonfuls into small, deep pans, and bake about forty minutes in moderate oven.

COCOA BUNS

- 2 tablespoonfuls of butter,
- 1/3 a cup of sugar,
- 1 egg,
- ¼ a teaspoonful of salt,
- 1 cup of scalded milk,
- 2 compressed yeast cakes softened in ½ a cup of warm water,
- ¼ a teaspoonful of extract cinnamon,
- ½ a cup of Baker's Breakfast Cocoa,
- 3 ½ to 4 cups of flour.

Mix in order given, having dough as soft as can be handled, turn onto moulding board, roll into a square about an inch in thickness, sprinkle on one-half cup of currants, fold the sides to meet the centre, then each end to centre, and fold again. Roll as at first, using another one-half cup currants, fold, roll and fold again. Place in a bowl which is set in pan of warm water, let raise forty minutes. Shape, place in pan, let raise until doubled in size. Bake fifteen to twenty minutes. As you take from oven, brush the top with white of one egg beaten with one-half cup confectioners' sugar. Let stand five minutes. Then they are ready to serve.

MRS. RORER'S CHOCOLATE CAKE

- 2 ounces of chocolate,
- 4 eggs,
- ½ a cup of milk,
- 1 teaspoonful of vanilla,
- ½ a cup of butter,
- 1 ½ cups of sugar,
- 1 heaping teaspoonful of baking powder,
- 1 ¾ cups of flour.

Dissolve the chocolate in five tablespoonfuls of boiling water. Beat the butter to a cream, add the yolks, beat again, then the milk, then the melted chocolate and flour. Give the whole a vigorous beating. Now beat the whites of the eggs to a stiff froth, and stir them carefully into the mixture; add the vanilla and baking powder. Mix quickly and lightly, turn into well-greased cake pan and bake in a moderate oven forty-five minutes. —*From Mrs. Rorer's Cook Book.*

MRS. LINCOLN'S CHOCOLATE CARAMELS

One cup of molasses, half a cup of sugar, one-quarter of a pound of chocolate cut fine, half a cup of milk, and one heaping table-spoonful of butter. Boil all together, stirring all the time. When it hardens in cold water, pour it into shallow pans, and as it cools cut in small squares.—*From Mrs. Lincoln's Boston Cook Book.*

MISS FARMER'S CHOCOLATE NOUGAT CAKE

- ¼ a cup of butter,
- 1 ½ cups of powdered sugar,
- 1 egg,
- 1 cup of milk,
- 2 cups of bread flour,
- 3 teaspoonfuls of baking powder,
- ½ teaspoonful of vanilla,
- 2 squares of chocolate, melted,
- ½ a cup of powdered sugar,
- 2/3 a cup of almonds blanched and shredded.

Cream the butter, add gradually one and one-half cups of sugar, and egg unbeaten; when well mixed, add two-thirds milk, flour mixed and sifted with baking powder, and vanilla. To melted chocolate add one-third a cup of powdered sugar, place on range, add gradually remaining milk, and cook until smooth. Cool slightly and add to cake mixture. Bake fifteen to twenty minutes in round layer-cake pans. Put between layers and on top of cake White Mountain Cream sprinkled with almonds. — *From Boston Cooking School Cook Book — Fannie Merritt Farmer.*

MRS. ARMSTRONG'S CHOCOLATE PUDDING

Soften three cups of stale bread in an equal quantity of milk. Melt two squares of Walter Baker & Co.'s Chocolate over hot water and mix with half a cup of sugar, a little salt, three beaten eggs and half a teaspoonful of vanilla. Mix this thoroughly with the bread and place in well-buttered custard-cups. Steam about half an hour (according to size) and serve in the cups or turned out on warm plate. — *Mrs. Helen Armstrong.*

MRS. ARMSTRONG'S CHOCOLATE CHAR-LOTTE

Soak a quarter of a package of gelatine in one-fourth of a cupful of cold water. Whip one pint of cream to a froth and put it in a bowl, which should be placed in a pan of ice water. Put an ounce of Walter Baker & Co.'s Chocolate in a small pan with two tablespoonfuls of sugar and one of boiling water, and stir over the hot fire until smooth and glossy. Add to this a gill of hot milk and the soaked gelatine, and stir until the gelatine is dissolved. Sprinkle a generous half cupful of powdered sugar over the cream. Now add the chocolate and gelatine mixture and stir gently until it begins to thicken. Line a quart charlotte mould with lady fingers, and when the cream is so thick that it will just pour, turn it gently into the mould. Place the charlotte in a cold place for an hour or more, and at serving time turn out on a flat dish. — *Mrs. Helen Armstrong.*

CHOCOLATE JELLY WITH CRYSTALLIZED GREEN GAGES

Dissolve in a quart of water three tablespoonfuls of grated chocolate; let come to a boil; simmer ten minutes; add a cup of sugar and a box of gelatine (that has been softened in a cup of water) and strain through a jelly bag or two thicknesses of cheese-cloth. When almost cold, add a dessertspoonful of vanilla and a tablespoonful of brandy. Then whisk well; add half a pound of crystallized green gages cut into small pieces; pour into a pretty mould, and when cold serve with whipped cream.

MRS. BEDFORD'S CHOCOLATE CRULLERS

Cream two tablespoonfuls of butter and one-half of a cupful of sugar; gradually add the beaten yolks of three eggs and one and one-half cupfuls more of sugar, one cupful of sour milk, one teaspoonful of vanilla, two ounces of chocolate grated and melted over hot water, one-third of a teaspoonful of soda dissolved in one-half of a teaspoonful of boiling water, the whites of the eggs whipped to a stiff froth, and sufficient sifted flour to make a soft dough. Roll out, cut into oblongs; divide each into three strips, leaving the dough united at one end. Braid loosely, pinch the ends together and cook until golden-brown in smoking-hot fat. — *Mrs. Cornelia C. Bedford.*

MRS. BEDFORD'S HOT COCOA SAUCE FOR ICE-CREAM

Boil together one and one-half cupfuls of water and one cupful of sugar for two minutes; add one tablespoonful of arrowroot dissolved in a little cold water, stir for a moment, then boil until clear. Add two tablespoonfuls of cocoa which has been dissolved in a little hot water and a tiny pinch of salt and boil three minutes longer. Take from the fire and add one teaspoonful of vanilla.—*Mrs. Cornelia C. Bedford.*

MRS. BEDFORD'S CHOCOLATE MACAROONS

Grate one-quarter of a pound of chocolate and mix one-quarter of a pound of sifted powdered sugar and one-quarter of a pound of blanched and ground almonds. Add a pinch of cinnamon and mix to a soft paste with eggs beaten until thick. Drop in half-teaspoonfuls on slightly buttered paper and bake in a moderate oven. Do not take from the paper until cold; then brush the under side with cold water, and the paper can be readily stripped off.—*Mrs. Cornelia C. Bedford.*

MRS. EWING'S CREAMY COCOA

Stir together in a saucepan half a cup of Walter Baker & Co.'s Breakfast Cocoa, half a cup of flour, half a cup of granulated sugar and half a teaspoonful of salt. Add gradually one quart of boiling water and let the mixture boil five minutes, stirring it constantly. Remove from the fire, add a quart of boiling milk, and serve. If desired a spoonful of whipped cream may be put in each cup before filling with the cocoa.

The proportions given will make delicious, creamy cocoa, sufficient to serve twelve persons. The flour should be sifted before it is measured. —*By Mrs. Emma P. Ewing, author of "The Art of Cookery."*

MRS. EWING'S CREAMY CHOCOLATE

Mix together half a cup of sifted flour, half a cup of granulated sugar and half a teaspoonful of salt. Put into a saucepan half a cup of Walter Baker & Co.'s Premium No. 1 Chocolate, finely shaved. Add one quart of boiling water, stir until dissolved, add the flour, sugar and salt, and boil gently, stirring constantly, five minutes. Then stir in a quart of boiling milk, and serve with or without whipped cream.—*Mrs. Emma P. Ewing, author of "The Art of Cookery."*

MRS. HILL'S COCOA FRAPPÉ

Mix half a pound of cocoa and three cupfuls of sugar; cook with two cupfuls of boiling water until smooth; add to three quarts and a half of milk scalded with cinnamon bark; cook for ten minutes. Beat in the beaten whites of two eggs mixed with a cupful of sugar and a pint of whipped cream. Cool, flavor with vanilla extract, and freeze. Serve in cups. Garnish with whipped cream. — *Janet McKenzie Hill — Ladies' Home Journal.*

MRS. HILL'S CHOCOLATE PUFFS

Stir a cupful of flour into a cupful of water and half a cupful of butter, boiling together; remove from fire, beat in an ounce of melted chocolate, and, one at a time, three large eggs. Shape with forcing bag and rose tube. Bake, cut off the tops and put into each cake a tablespoonful of strawberry preserves. Cover with whipped cream sweetened and flavored. —*Janet McKenzie Hill — Ladies' Home Journal.*

MISS FARMER'S CHOCOLATE CREAM CANDY

- 2 cups of sugar,
- 2/3 a cup of milk,
- 1 tablespoonful of butter,
- 2 squares of chocolate,
- 1 teaspoonful of vanilla.

Put butter into granite saucepan; when melted add sugar and milk. Heat to boiling point; then add chocolate, and stir constantly until chocolate is melted. Boil thirteen minutes, remove from fire, add vanilla, and beat until creamy and mixture begins to sugar slightly around edge of saucepan. Pour at once into a buttered pan, cool slightly and mark in squares. Omit vanilla, and add, while cooking, one-fourth of a teaspoonful of cinnamon.—*Boston Cooking School Cook Book—Fannie Merritt Farmer.*

MRS. SALZBACHER'S CHOCOLATE HEARTS

Melt, by standing over hot water, three ounces of unsweetened chocolate; add a pound of sifted powdered sugar and mix thoroughly; work to a stiff yet pliable paste with the unbeaten whites of three eggs (or less), adding vanilla to flavor. If the paste seems too soft, add more sugar. Break off in small pieces and roll out about one-fourth of an inch thick, sprinkling the board and paste with granulated sugar instead of flour. Cut with a tiny heart-shaped cake cutter (any other small cake cutter will do), and place on pans oiled just enough to prevent sticking. Bake in a very moderate oven. When done, they will feel firm to the touch, a solid crust having formed over the top. They should be very light, and will loosen easily from the pan after being allowed to stand a moment to cool. The success of these cakes depends upon the oven, which should not be as cool as for meringue, nor quite so hot as for sponge cake. If properly made, they are very excellent and but little labor. Use the yolks for chocolate whips. — *From "Good Housekeeping."*

COCOA CHARLOTTE (Without Cream)

- 1 pint of water,
- Whites of 2 eggs,
- ½ a teaspoonful of vanilla,
- ½ a cup of sugar,
- 2 level tablespoonfuls of cornstarch,
- ½ a teaspoonful of cinnamon,
- 3 tablespoonfuls of cocoa.

Dissolve the cornstarch in a quarter of a cup of cold water, add it to the pint of boiling water, stir until it thickens, add the sugar and the cocoa, which have been thoroughly mixed together. Remove from the fire, add the cinnamon and vanilla, and pour slowly over the stiffly beaten whites of eggs. Pour at once into a pudding mould, and put away in a cold place to harden. Serve with plain cream. — *Mabel Richards Dulon.*

CHOCOLATE FUDGE WITH FRUIT

Two cups of sugar, one-half cup of milk, one-half cup of molasses, one-half cup of butter; mix all together and boil seven minutes; add one-half cup of Baker's Chocolate and boil seven minutes longer. Then add two tablespoonfuls of figs, two tablespoonfuls of raisins, one-half a cup of English walnuts and one teaspoonful of vanilla.

CHOCOLATE MACAROONS

Stir to a paste whites of seven eggs, three-fourths a pound of sifted sugar, one-half a pound of almonds pounded very fine, and two ounces of grated Baker's Chocolate. Have ready wafer paper cut round, on which lay pieces of the mixture rolled to fit the wafer. Press one-half a blanched almond on each macaroon and bake in a moderate oven.

PETITS FOUR

Bake a simple, light sponge cake in a shallow biscuit tin or dripping pan, and when cold turn out on the moulding board and cut into small dominoes or diamonds. They should be about an inch in depth. Split each one and spread jelly or frosting between the layers, then ice tops and sides with different tinted icings, pale green flavored with pistachio, pale pink with rose, yellow with orange, white with almond. Little domino cakes are also pretty. Ice the cakes on top and sides with white icing, then when hard put on a second layer of chocolate, using *Walter Baker & Co.'s Unsweetened Chocolate* and made as for layer cake, dipping the brush in the melted chocolate to make the spots.

Candied violets, bits of citron cut in fancy shapes, candied cherries and angelica may all be utilized in making pretty designs in decoration.—*American Housekeeper.*

POTATO CAKE

Two cups of white sugar, one cup of butter, one cup of hot mashed potatoes, one cup of chopped walnuts, half a cup of sweet milk, two cups of flour, four eggs well beaten, five teaspoonfuls of melted chocolate, one tablespoonful each of cloves, cinnamon and nutmeg, two teaspoonfuls of baking powder. Bake in layers and use marshmallow filling.

SPANISH CHOCOLATE CAKE

One cup of sugar, one-half a cup of butter, one-half a cup of sweet milk, three cups of flour, two eggs, one teaspoonful of soda dissolved in hot water. Put on the stove one cup of milk, one-half a cup of Baker's Chocolate, grated; stir until dissolved; then stir into it one cup of sugar and the yolk of one egg stirred together; when cool flavor with vanilla. While this is cooling beat up the first part of the cake and add the chocolate custard. Bake in layers. Ice on top and between the layers.

PEPPERMINTS, CHOCOLATE MINTS, Etc.

(Uncooked Fondant)

- White of 1 egg,
- 2 tablespoonfuls of cold water,
- Sifted confectioner's sugar,
- ½ teaspoonful of essence of peppermint or a few drops of oil of peppermint,
- 1 or 2 squares of Baker's Chocolate,
- Green color paste,
- Pink color paste.

Beat the egg on a plate, add the cold water and gradually work in sugar enough to make a firm paste. Divide the sugar paste into three parts. To one part add the peppermint and a very little of the green color paste. Take the paste from the jar with a wooden tooth pick, add but a little. Work and knead the mixture until the paste is evenly distributed throughout. Roll the candy into a sheet one-fourth an inch thick, then cut out into small rounds or other shape with any utensil that is convenient. Color the second part a very delicate pink, flavor with rose extract and cut out in the same manner as the first. To the last part add one or two squares of Baker's Chocolate, melted over hot water, and flavor with peppermint. Add also a little water, as the chocolate will make the mixture thick and crumbly. Begin by adding a tablespoonful of water, then add more if necessary, knead and cut these as the others.

CHOCOLATE CARAMEL WALNUTS

(Uncooked Fondant)

- White of 1 egg,
- 3 tablespoonfuls of maple or caramel syrup,
- 1 tablespoonful of water,
- Sifted confectioner's sugar,
- 1 teaspoonful of vanilla extract,
- 2 or more squares of Baker's Chocolate,
- English walnuts.

Beat the white of egg slightly, add the syrup, water, sugar as needed, the chocolate, melted over hot water, and the vanilla, also more water if necessary. Work with a silver plated knife and knead until thoroughly mixed, then break off small pieces of uniform size and roll them into balls, in the hollow of the hand, flatten the balls a little, set the half of an English walnut upon each, pressing the nut into the candy and thus flattening it still more. The caramel gives the chocolate a particularly nice flavor.

HOW TO COAT CANDIES, &c., WITH BAKER'S "DOT" CHOCOLATE

Half a pound of "Dot" Chocolate will coat quite a number of candy or other "centers," but as depth of chocolate and an even temperature during the whole time one is at work are essential, it is well, when convenient, to melt a larger quantity of chocolate. When cold, the unused chocolate may be cut from the dish and set aside for use at a future time. If the chocolate be at the proper temperature when the centers are dipped in it, it will give a rich, glossy coating free from spots, and the candies will not have a spreading base. After a few centers have been dipped set them in a cool place to harden. The necessary utensils are a wire fork and a very small double boiler. The inner dish of the boiler should be of such size that the melted chocolate will come nearly to the top of it. Break the chocolate in small pieces and surround with warm water, stir occasionally while melting. When the melted chocolate has cooled to about 80° F. it is ready to use. Drop whatever is to be coated into the chocolate, with the fork push it below the chocolate, lift out, draw across the edge of the dish and drop onto a piece of table oil cloth or onto waxed paper. Do not let a drop of water get into the chocolate.

CHOCOLATE DIPPED PEPPERMINTS

(Uncooked Fondant)

Prepare green, white, pink and chocolate colored mints by the first recipe. After they have dried off a little run a spatula under each and turn to dry the other side. Coat with Baker's "Dot" Chocolate.

GINGER, CHERRY, APRICOT and NUT CHOCO-LATES

- White of 1 egg,
- 2 tablespoonfuls of cold water,
- Sifted confectioner's sugar,
- Almond or rose extract,
- Preserved ginger,
- Candied cherries,
- Candied apricots,
- Halves of almond,
- Halves of pecan nuts,
- ½ a pound of Baker's "Dot" Chocolate.

Use the first four ingredients in making uncooked fondant. (Caramel syrup is a great addition to this fondant, especially if nuts are to be used. Use three tablespoonfuls of syrup and one tablespoonful of water with one egg white instead of the two tablespoonfuls of water indicated in the recipe). Work the fondant for some time, then break off little bits and wrap around small pieces of the fruit, then roll in the hollow of the hand into balls or oblongs. For other candies, roll a piece of the fondant into a ball, flatten it with the fingers and use to cover a whole pecan or English walnut meat. Set each shape on a plate as it is finished. They will harden very quickly. Dip these, one by one, in Baker's "Dot" Chocolate and set on an oil cloth.

CHOCOLATE PEANUT CLUSTERS

Shell a quart of freshly-roasted peanuts and remove the skins. Drop the peanuts, one by one, into the center of a dish of "Dot" Chocolate made ready for use; lift out onto oil cloth with a dipping fork (a wire fork comes for the purpose, but a silver oyster fork answers nicely) to make groups of three nuts, — two below, side by side, and one above and between the others.

CHOCOLATE COATED ALMONDS

Select nuts that are plump at the ends. Use them without blanching. Brush, to remove dust. Melt "Dot" Chocolate and when cooled properly drop the nuts, one at a time, into the center of it; push the nuts under with the fork, then drop onto waxed paper or oil cloth. In removing the fork make a design on the top of each nut. These are easily prepared and are particularly good.

PLAIN AND CHOCOLATE DIPPED PARISIAN SWEETS

- ½ a cup of Sultana raisins,
- 5 figs,
- 1 cup of dates,
- 2 ounces citron,
- 2/3 a cup of nut meats, (almonds, filberts, pecans or walnuts, one variety or a mixture),
- 1 ½ ounces of Baker's Chocolate,
- 1/3 a cup of confectioner's sugar,
- ¼ a teaspoonful of salt,
- Chocolate Fondant or
- Baker's "Dot" Chocolate.

Pour boiling water over the figs and dates, let boil up once, then drain as dry as possible; remove stones from the dates, the stem ends from the figs; chop the fruit and nut meats (almonds should be blanched) in a food chopper; add the salt; and the sugar and work the whole to a smooth paste; add the chocolate, melted, and work it evenly through the mass. Add more sugar if it is needed and roll the mixture into a sheet one-fourth an inch thick. Cut into strips an inch wide. Cut the strips into diamond-shaped pieces (or squares); roll these in confectioner's sugar or dip them in chocolate fondant or in Baker's "Dot" Chocolate, and sprinkle a little fine-chopped pistachio nut meats on the top of the dipped pieces. When rolling the mixture use confectioner's sugar on board and rolling pin.

STUFFED DATES, CHOCOLATE DIPPED

Cut choice dates open on one side and remove the seeds. Fill the open space in the dates with a strip of preserved ginger or pineapple, chopped nuts or chopped nuts mixed with white or chocolate fondant; press the dates into a compact form to keep in the filling, then dip them, one by one, in "Dot" Chocolate.

CHOCOLATE OYSTERETTES, PLAIN AND WITH CHOPPED FIGS

- Oyster crackers, salted preferred, fine-chopped, roasted peanuts or raisins or 3 or 4 basket figs or a little French fruit cut in very small bits,
- ½ a pound or more of Baker's "Dot" Chocolate.

Select fresh-baked crackers free from crumbs. Dip in "Dot" Chocolate, made ready as in previous recipes, and dispose on oil cloth or waxed paper. For a change add figs or other fruit, cut very fine, or chopped nuts to the chocolate ready for dipping.

TURKISH PASTE WITH FRENCH FRUIT, CHOC-OLATE FLAVORED

- 3 level tablespoonfuls of granulated gelatine,
- ½ a cup of cold water,
- 2 cups of sugar,
- 2/3 a cup of cold water,
- 1 teaspoonful of ground cinnamon,
- 2 squares of Baker's Chocolate,
- 1 teaspoonful of vanilla extract,
- 1 cup of French candied fruit, cherries, angelica, citron, etc., chopped fine.

Let the gelatine stand in the half cup of cold water until it has taken up all of the water. Stir the sugar and the two-thirds a cup of cold water over the fire until the sugar is dissolved and the syrup is boiling, then add the gelatine and let cook twenty minutes; add the cinnamon, the chocolate, melted over hot water, and beat all together, then add the vanilla and the fruit; let stand in a cool place for a time, then when it thickens a little turn into an *un*buttered bread pan and set aside until the next day. To unmold separate the paste from the pan—at the edge—with a sharp-pointed knife. Sift confectioner's sugar over the top, then with the tips of the fingers gently pull the paste from the pan to a board dredged with confectioner's sugar; cut into strips, then into small squares. Roll each square in confectioner's sugar. In cutting keep sugar between the knife and the paste.

CHOICE CHOCOLATE PECAN PRALINES

- 3 cups of granulated sugar,
- 1 cup of cream,
- 1 cup of sugar cooked to caramel,
- 2 squares of Baker's Chocolate,
- 3 cups of pecan nut meats.

Stir the sugar and cream over the fire until the sugar is melted, then let boil to the soft ball degree, or to 236° F. Add the chocolate, melted or shaved fine, and beat it in, then pour the mixture onto the cup of sugar cooked to caramel; let the mixture boil up once, then remove from the fire; add the nut meats and beat until the mass begins to thicken. When cold enough to hold its shape drop onto an oil cloth or marble, a teaspoonful in a place, and at once set a half nut meat on each. Two persons are needed to make these pralines, one to drop the mixture, the other to decorate with the halves of the nuts. The mixture becomes smooth and firm almost instantly. Maple or brown sugar may be used in place of all or a part of the quantity of granulated sugar designated.

VASSAR FUDGE

- 2 cups of white granulated sugar,
- 1 cup of cream,
- 1 tablespoonful of butter,
- ¼ a cake of Baker's Premium No. 1 Chocolate.

Put in the sugar and cream, and when this becomes hot put in the chocolate, broken up into fine pieces. Stir vigorously and constantly. Put in butter when it begins to boil. Stir until it creams when beaten on a saucer. Then remove and beat until quite cool and pour into buttered tins. When cold cut in diamond-shaped pieces.

SMITH COLLEGE FUDGE

Melt one-quarter cup of butter. Mix together in a separate dish one cup of white sugar, one cup of brown sugar, one-quarter cup of molasses and one-half cup of cream. Add this to the butter, and after it has been brought to a boil continue boiling for two and one-half minutes, stirring rapidly. Then add two squares of Baker's Premium No. 1 Chocolate, scraped fine. Boil this five minutes, stirring it first rapidly, and then more slowly towards the end. After it has been taken from the fire, add one and one-half teaspoonfuls of vanilla. Then stir constantly until the mass thickens. Pour into buttered pan and set in a cool place.

WELLESLEY MARSHMALLOW FUDGE

Heat two cups of granulated sugar and one cup of rich milk (cream is better). Add two squares of Baker's Chocolate, and boil until it hardens in cold water. Just before it is done add a small piece of butter, then begin to stir in marshmallows, crushing and beating them with a spoon. Continue to stir in marshmallows, after the fudge has been taken from the fire, until half a pound has been stirred into the fudge. Cool in sheets three-quarters of an inch thick, and cut in cubes.

DOUBLE FUDGE

- 2 cups of granulated sugar,
- 2 squares of Baker's Chocolate,
- ½ a cup of cream,
- 1 tablespoonful of butter.

Boil seven minutes; then beat and spread in buttered tin to cool.

- 2 cups of brown sugar,
- ½ a cup of cream,
- 1 teaspoonful of vanilla extract,
- 1 cup of walnut meats, chopped fine,
- Butter size of a walnut.

Boil ten minutes; then beat and pour on top of fudge already in pan. When cool, cut in squares.

MARBLED FUDGE

- 2 cups of granulated sugar,
- ¼ a cup of glucose (pure corn syrup),
- 1 ½ cups of cream,
- 1 tablespoonful of butter,
- 2 squares of Baker's Chocolate, scraped fine or melted,
- 2 teaspoonfuls of vanilla.

Stir the sugar, glucose and cream over a slack fire until the sugar is melted; move the saucepan to a hotter part of the range and continue stirring until the mixture boils, then let boil, stirring every three or four minutes very gently, until the thermometer registers 236° F., or, till a soft ball can be formed in cold water. Remove from the fire and pour one-half of the mixture over the chocolate. Set both dishes on a cake rack, or on something that will allow the air to circulate below the dishes. When the mixture cools a little, get some one to beat one dish of the fudge; add a teaspoonful of vanilla to each dish, and beat until thick and slightly grainy, then put the mixture in a pan, lined with waxed paper, first a little of one and then of the other, to give a marbled effect. When nearly cold turn from the pan, peel off the paper and cut into cubes.

FUDGE HEARTS OR ROUNDS

- 2 cups of granulated sugar,
- 1/3 a cup of condensed milk,
- 1/3 a cup of water,
- ¼ a cup of butter,
- 1 ½ squares of Baker's Chocolate,
- 1 teaspoonful of vanilla extract.

Boil the sugar, milk and water to 236° F., or to the "soft ball" degree; stir gently every few minutes; add the butter and let boil up vigorously, then remove from the fire and add the chocolate; let stand undisturbed until cool, then add the vanilla and beat the candy until it thickens and begins to sugar. Pour into a pan lined with paper to stand until cooled somewhat; turn from the mold and with a French cutter or a sharp edged tube cut into symmetrical shapes.

MARSHMALLOW FUDGE

1st BATCH

- 2 cups of granulated sugar,
- 1 cup of cream,
- ¼ a teaspoonful of salt,
- 1 tablespoonful of butter,
- 2 squares of Baker's Chocolate,
- 1 teaspoonful of vanilla,
- Nearly half a pound of marshmallows, split in halves.

2nd BATCH

- 2 cups of granulated sugar,
- 1 cup of cream,
- ¼ a teaspoonful of salt,
- 1 tablespoonful of butter,
- 2 squares of Baker's Chocolate,
- 1 teaspoonful of vanilla.

Start with the first batch and when this is nearly boiled enough, set the second batch to cook, preparing it in the same manner as the first. Stir the sugar and cream, over a rather slack fire, until the sugar is melted, when the sugar boils wash down the sides of the pan as in making fondant, set in the thermometer and cook over a quick fire, without stirring, to the soft ball degree, 236° F.; add the butter, salt and chocolate, melted or shaved fine, and let boil up vigorously, then remove to a cake cooler (or two spoon handles to allow a circulation of air below the pan). In the meantime the second batch should be cooking and the marshmallows be gotten ready. When the first batch is about cold add the vanilla and beat the candy vigorously until it begins to thicken, then turn it into a pan lined with waxed paper. At once dispose the halves of marshmallows close together upon the top of the fudge. Soon the other dish of fudge will be ready; set it into cold water and when nearly cold, add the vanilla and beat as in the first batch, then pour it over the marshmallows. When the whole is about cold turn it onto a marble, or hardwood board, pull off the paper and cut into cubes. If one is able to work very quickly, but one batch need be prepared, half of it being spread over the marshmallows.

CHOCOLATE DIPPED FRUIT FUDGE

FRUIT FUDGE

- 1 ½ cups of granulated sugar,
- 1 cup of Maple Syrup,
- 1 ½ cups of glucose (pure corn syrup),
- ½ a cup of thick cream, or
- 1/3 a cup of milk and 1/4 a cup of butter,
- ¾ a cup of fruit, figs, and candied cherries and apricots, cut in small pieces.

CHOCOLATE FOR DIPPING

- ½ a cake or more of Baker's "Dot" Chocolate.

Stir the sugar, syrup, glucose and cream until the sugar is melted, cover and let boil three or four minutes, then uncover and let boil stirring often but very gently until a soft ball may be formed in cold water, or, until the thermometer registers 236° F. Set the saucepan on a cake cooler and when the mixture becomes cool, add the fruit and beat until it becomes thick, then turn into pans lined with waxed paper. In about fifteen minutes cut into squares. Coat these with the "Dot" Chocolate.

CHOCOLATE COCOANUT CAKES

- 2/3 a cup of granulated sugar,
- ¼ a cup scant measure of water,
- One cup, less one tablespoonful, of glucose,
- ½ a pound of dessicated cocoanut,
- ½ a pound or Baker's "Dot" Chocolate.

Heat the sugar, water and glucose to the boiling point, add the cocoanut and stir constantly while cooking to the soft ball degree, or, until a little of the candy dropped on a cold marble may be rolled into a ball. Drop, by small teaspoonfuls, onto a marble or waxed paper, to make small, thick, rather uneven rounds. When cold coat with "Dot" Chocolate melted over hot water and cooled properly. These cakes are very easily coated.

BAKER'S CHOCOLATE "DIVINITY"

- 1 ½ cups of brown sugar,
- 1 cup of maple syrup,
- ½ a cup of glucose pure corn syrup,
- ½ a cup of water,
- ¼ a teaspoonful of salt,
- The whites of 2 eggs,
- 1 cup of nut meats, chopped fine,
- 2 squares of Baker's Chocolate, broken in pieces.

Let the sugar, syrup, glucose and water stand on the back of the range, stirring occasionally, until the sugar is melted, then cover and let boil five minutes. Remove the cover and let boil to soft crack, 287° F., or, until when tested in water a ball that rattles in the cup will be formed. Add the salt and chocolate and beat over the fire, until the chocolate is melted, then pour in a fine stream onto the whites of eggs, beaten dry, beating constantly meanwhile; add the nuts and pour into a pan lined with waxed paper. In about fifteen minutes lift the candy from the pan (by the ends of the paper left for the purpose) and cut it into small oblongs or squares. The candy must be stirred constantly during the last of the cooking. In cooking without a thermometer one is liable to remove the candy from the fire too soon—if this happens, return, egg whites and all, to the saucepan, set this into a dish of boiling water and stir constantly until the mixture thickens, then pour into the pan lined with paper. On no account let even a few drops of water boil into the candy.

CHOCOLATE NOUGATINES

- 1 cup of granulated sugar,
- ½ a cup of glucose,
- ½ a cup of honey (strained),
- Piece of paraffine size of a pea,
- ¼ a cup of water,
- ¼ a teaspoonful of salt,
- The whites of 2 eggs, beaten dry,
- 1 cup of almond or English walnut meats, chopped fine,
- 1 teaspoonful of vanilla,
- About ½ a pound of Baker's "Dot" Chocolate.

Put the sugar, glucose, honey, paraffine and water over the fire, stir occasionally and let boil to the hard ball degree, about 248° F. Add the salt to the eggs before beating them, and gradually pour on part of the syrup, beating constantly meanwhile with the egg beater; return the rest of the syrup to the fire and let boil until it is brittle when tested in cold water or to 290° F. Then turn this gradually onto the eggs, beating constantly meanwhile. Return the whole to the saucepan, set over the fire on an asbestos mat and beat constantly until it becomes crisp when tested in cold water. Pour into a buttered pan a little larger than an ordinary bread pan and set aside to become cold. When cold cut into pieces about an inch and a quarter

long and three-eighths of an inch wide and thick. Coat these with
"Dot" Chocolate.

PLAIN CHOCOLATE CARAMELS

- 2 ½ cups of sugar,
- ¾ cup of glucose, (pure corn syrup),
- ½ a cup of butter,
- 1/8 a teaspoonful of cream of tartar,
- 2 ½ cups of whole milk, (not skimmed),
- 2 ½ squares of Baker's Chocolate,
- 1 teaspoonful of vanilla extract.

Put the sugar, glucose, butter, cream of tartar and one cup of the milk over the fire, stir constantly, and when the mass has boiled a few moments, gradually stir in the rest of the milk. Do not let the mixture stop boiling while the milk is being added. Stir every few moments and cook to 248° F., or, until when tested in cold water, a hard ball may be formed; add the chocolate and vanilla and beat them thoroughly through the candy, then turn it into two bread pans. When nearly cold cut into squares.

CHOCOLATE NUT CARAMELS

- 2 cups of granulated sugar,
- 1 ½ cups of glucose (pure corn syrup),
- 2 cups of cream,
- 1 cup of butter,
- 3 or 4 squares of Baker's Chocolate,
- 1 ½ cups of English walnut meats,
- 2 teaspoonfuls of vanilla extract.

Put the sugar, glucose, *one* cup of the cream and the butter over the fire; stir and cook until the mixture boils vigorously, then gradually add the other cup of cream. Do not allow the mixture to stop boiling while the cream is being added. Cook until the thermometer registers 250° F., stirring gently—move the thermometer, to stir beneath it—every four or five minutes. Without a thermometer boil until—when tested by dropping a little in cold water—a hard ball may be formed in the water. Remove from the fire, add the chocolate and nuts and beat until the chocolate is melted; beat in the vanilla and turn into a biscuit pan, nicely oiled or buttered, to make a sheet three-fourths an inch thick. When nearly cold turn from the pan and cut into cubes.

RIBBON CARAMELS

CHOCOLATE LAYERS

- 1 ¼ cups of granulated sugar,
- ½ cup of glucose (pure corn syrup) *scant* measure,
- ¼ a cup of butter,
- 1/16 a teaspoonful of cream of tartar,
- 1 ¼ cups of rich milk,
- 1 ¼ squares of Baker's Premium Chocolate,
- 1 teaspoonful of vanilla extract.

WHITE LAYER

- 2/3 a cup of granulated sugar,
- ¼ (scant) a cup of water,
- 1 cup, less one tablespoonful, of glucose (pure corn syrup),
- 1/3 a pound of dessicated cocoanut.

Put the sugar, glucose, butter, cream of tartar and the fourth a cup of milk over the fire, stir until the mixture boils, then very gradually stir in the rest of the milk. Let cook, stirring occasionally, to 248° F., or until, when tested in water or on a cold marble, a pretty firm ball may be formed. Add the chocolate and vanilla, mix thoroughly and turn into two well-buttered shallow pans. For the white layer, put the sugar, water and glucose over the fire, stir until boiling, then add the cocoanut and stir occasionally until a soft ball may be formed when a little of the mixture is dropped upon a cold marble.

Put this mixture over the fire, to dissolve the sugar, but do not let it begin to boil until the chocolate layers are turned into the pans. When the white mixture is ready, turn enough of it onto one of the chocolate layers to make a layer about one-third an inch thick. Have the other chocolate layer cooled, by standing in cold water; remove it from the pan and dispose above the cocoanut layer. Let stand until cold and firm, then cut in cubes; wrap each cube in waxed paper.

FONDANT

- 4 cups of granulated sugar,
- 1 ½ cups of cold water,
- ¼ a teaspoonful of cream of tartar, or 3 drops of acetic acid.

Stir the sugar and water in a saucepan, set on the back part of the range, until the sugar is melted, then draw the saucepan to a hotter part of the range, and stir until the boiling point is reached; add the cream of tartar or acid and, with the hand or a cloth wet repeatedly in cold water, wash down the sides of the saucepan, to remove any grains of sugar that have been thrown there. Cover the saucepan and let boil rapidly three or four minutes. Remove the cover, set in the thermometer—if one is to be used—and let cook very rapidly to 240° F., or the soft ball degree. Wet the hand in cold water and with it dampen a marble slab or a large platter, then without jarring the syrup turn it onto the marble or platter. Do not scrape out the saucepan or allow the last of the syrup to drip from it, as sugary portions will spoil the fondant by making it grainy. When the syrup is cold, with a metal scraper or a wooden spatula, turn the edges of the mass towards the center, and continue turning the edges in until the mass begins to thicken and grow white, then work it up into a ball, scraping all the sugar from the marble onto the mass; knead slightly, then cover closely with a heavy piece of cotton cloth wrung out of cold water. Let the sugar stand for an hour or longer to ripen, then remove the damp cloth and cut the mass into pieces; press these closely into a kitchen bowl, cover with a cloth wrung out of water (this cloth must not touch the fondant) and then with heavy paper. The fondant may be used the next day, but is in better condition after several days, and may be kept almost indefinitely, if the cloth covering it be wrung out of cold water and replaced once in five or six days. Fondant may be used, white or delicately colored with vegetable color-pastes or with chocolate, as frosting for small cakes, or éclairs or for making candy "centers," to be coated with chocolate or with some of the same fondant tinted and flavored appropriately.

ALMOND CHOCOLATE CREAMS

CENTERS

- ¼ a cup of blanched almonds, chopped fine,
- ½ a cup of fondant,
- ¼ a teaspoonful of vanilla,
- Confectioner's sugar for kneading and shaping.

CHOCOLATE COATING

- About 1 cup of fondant,
- 2 squares of Baker's Chocolate,
- 1 teaspoonful of vanilla extract,
- Few drops of water, as needed,
- Halves of blanched almonds.

Mix the chopped almonds with the fondant and vanilla; add confectioner's sugar, a little at a time, and knead the mass thoroughly, on a marble or large platter; shape into a long roll, then cut into small pieces of the same size. Shape these into balls a generous half inch in diameter and leave them about an hour to harden on the outside. Put the fondant for the coating and the chocolate (shaved or broken in pieces) in a double boiler (with hot water in the lower receptacle); add the vanilla and the water and heat until melted; take out the spoon and put in a dipping fork (a wire fork costing about ten cents) beat the fondant, to keep it from crusting and drop in a "center;" with the fork cover it with fondant; put the fork under it and lift it out, scrape the fork lightly on the edge of the dish, to remove superfluous candy, turn the fork over and drop the bon-bon onto waxed paper. Make a design with the fork in taking it from the candy. At once press half of a blanched almond on the top of the candy, or the design made with the fork will suffice. If at any time the coating be too thick, add a few drops of water. If any be left over, use it to coat whole nuts or cherries.

CHERRY CHOCOLATE CREAMS

CENTERS

- ¼ a cup of candied cherries, chopped fine,
- ½ a cup of fondant.

CHOCOLATE COATING

- About one cup of fondant,
- 2 squares of Baker's Chocolate,
- 1 teaspoonful of vanilla extract,
- Bits of cherry.

Prepare the centers and coat in the same manner as the almond creams.

CHOCOLATE PEPPERMINTS

Melt a little fondant and flavor it to taste with essence of peppermint; leave the mixture white or tint very delicately with green or pink color-paste. With a teaspoon drop the mixture onto waxed paper to make rounds of the same size—about one inch and a quarter in diameter—let these stand in a cool place about one hour. Put about a cup of fondant in a double boiler, add two ounces of chocolate and a teaspoonful of boiling water, then stir (over hot water) until the fondant and chocolate are melted and evenly mixed together; then drop the peppermints, one by one, into the chocolate mixture, and remove them with the fork to a piece of oil cloth; let stand until the chocolate is set, when they are ready to use.

FIG-AND-NUT CHOCOLATES

- 5 figs,
- 3 or 4 tablespoonfuls of water or sherry wine,
- ½ a cup of English walnut meats,
- Powdered sugar,
- Fondant,
- 3 or 4 ounces of Baker's Chocolate,
- 1 teaspoonful of vanilla.

Remove the stem and hard place around the blossom end of the figs, and let steam, with the water or wine, in a double boiler until softened, then add the nuts and chop very fine. Add powdered sugar as is needed to shape the mixture into balls. Melt the chocolate, using enough to secure the shade of brown desired in the coating and add to the fondant with the vanilla. Coat the fig-and-nut balls and drop them with the fork onto a piece of oil cloth or waxed paper in the same manner as the cherry bon-bons. These may be dipped in "Dot" Chocolate instead of fondant.

CHOCOLATE MARSHMALLOWS

Cut the marshmallows in halves, and put them, one by one, cut side down, in chocolate fondant (as prepared for almond and cherry chocolate creams), melted over hot water and flavored to taste with vanilla. Beat the chocolate with the fork, that it may not crust over, lift out the marshmallow, turn it and, in removing the fork, leave its imprint in the chocolate; sprinkle at once with a little fine-chopped pistachio nut meat. To prepare the nuts, set them over the fire in tepid water to cover, heat to the boiling point, drain, cover with cold water, then take them up, one by one, and with the thumb and finger push the meat from the skin.

MAPLE FONDANT ACORNS

- 2 cups of maple syrup,
- 1 ¾ cups of granulated sugar,
- ¾ a cup of cold water,
- Confectioner's sugar,
- 2 or more squares of Baker's Chocolate,
- 1 teaspoon of vanilla,
- About ¼ a cup of fine-chopped almonds, browned in the oven.

Make fondant of the syrup, granulated sugar and cold water, following the directions given for fondant made of granulated sugar (cream of tartar or other acid is not required in maple fondant). Work some of the fondant, adding confectioner's sugar as needed, into cone shapes; let these stand an hour or longer to harden upon the outside. Put a little of the fondant in a dish over hot water; add Baker's Chocolate and vanilla as desired and beat till the chocolate is evenly mixed through the fondant, then dip the cones in the chocolate and set them on a piece of oil cloth or waxed paper. When all are dipped, lift the first one dipped from the paper and dip the base again in the chocolate, and then in the chopped-and-browned almonds. Continue until all have been dipped.

CHOCOLATE ALMOND BARS

- ½ a cup of sugar,
- ¾ a cup of glucose,
- ½ a cup of water,
- (¼ an ounce of paraffine at discretion),
- ½ a cup of blanched almonds, chopped fine,
- 1/3 the recipe for fondant,
- 3 or 4 ozs. of Baker's Chocolate,
- 1 teaspoonful of vanilla.

Melt the sugar in the water and glucose and let boil to about 252° F., or between a soft and a hard ball. Without the paraffine cook a little higher than with it. Add the almonds and the vanilla, mix thoroughly and turn onto a marble or platter over which powdered sugar has been sifted. Turn out the candy in such a way that it will take a rectangular shape on the marble. When cool enough score it in strips about an inch and a quarter wide, and, as it grows cooler, lift the strips, one by one, to a board and cut them in pieces half or three-quarters of an inch wide. When cold, drop them, sugar side down, in chocolate fondant prepared for "dipping." With the fork push them below the fondant, lift out, drain as much as possible, and set onto oil cloth. These improve upon keeping.

ALMOND FONDANT STICKS

- 2 ½ cups of coffee A or granulated sugar,
- ¼ a cup of glucose,
- ½ a cup of water,
- ¼ a pound of almond paste,
- ¼ a pound of Baker's Premium Chocolate,
- 1 teaspoonful of vanilla extract,
- ½ a pound of Baker's "Dot" Chocolate.

Put the sugar, glucose and water over the fire. Stir until the sugar is dissolved. Wash down the sides of the kettle as in making fondant. Let boil to the soft ball degree or to 238° F. Add the almond paste, cut into small, thin pieces, let boil up vigorously, then turn onto a damp marble. When nearly cold turn to a cream with a wooden spatula. It will take considerable time to turn this mixture to fondant. Cover and let stand half an hour. Add the Baker's Premium Chocolate, melted over hot water, and knead it in thoroughly. Add at the same time the vanilla. The chocolate must be added warm. At once cut off a portion of the fondant and knead it into a round ball; then roll it lightly under the fingers into a long strip the shape and size of a lead pencil; form as many of these strips as desired; cut the strips into two-inch lengths and let stand to become firm. Have ready the "Dot" Chocolate melted over hot water and in this coat the prepared sticks leaving the surface a little rough.

ALMOND FONDANT BALLS

Roll part of the almond fondant into small balls. Some of the "Dot" Chocolate will be left after dipping the almond chocolate sticks. Remelt this over hot water, and in it coat the balls lightly. As each ball is coated with the chocolate drop it onto a plate of chopped pistachio nut meats or of chopped cocoanut (fresh or dessicated). With a spoon sprinkle the chopped material over the balls.

WALNUT CREAM-CHOCOLATES

- 2 ½ cups of granulated sugar,
- ½ a cup of condensed milk,
- ½ a cup of water,
- 3 or 4 tablespoonfuls of thick caramel syrup,
- A little water,
- 1 teaspoonful of vanilla,
- ½ a pound of Baker's "Dot" Chocolate.

Put the sugar, condensed milk and water over the fire to boil, stir gently but often, and let cook to the soft ball stage, or to 238°F. Pour on a damp marble and let stand undisturbed until cold; turn to a cream, then gather into a compact mass; cover with a bowl and let stand for thirty minutes; then knead the cream; put it into a double boiler; add the caramel syrup and the vanilla; stir constantly while the mixture becomes warm and thin; add a tablespoonful or two of water, if necessary, and drop the cream mixture into impressions made in cornstarch. Use two teaspoons to drop the cream. When the candy is cold, pick it from the starch. With a small brush remove the starch that sticks to the candy shapes. Coat each piece with "Dot" Chocolate. As each piece is coated and dropped onto the oil cloth, set half an English walnut meat upon the top.

TO MOLD CANDY IN STARCH IMPRESSIONS

Many candies, especially such as are of some variety of fondant, are thin when warm and solidify on the outside when cold, so that they may be "dipped" or coated with chocolate. To shape candy of this sort, fill a low pan with cornstarch, making it smooth upon the top. Have ready molds made of plaster paris, glued to a thin strip of wood, press these into the cornstarch; lift from the starch and repeat the impressions as many times as the space allows. If molds are not available a thimble, round piece of wood, or the stopper of an oil or vinegar cruet will answer the purpose, though the impressions must be made one at a time.

CHOCOLATE BUTTER CREAMS

- 2 ½ cups of sugar,
- ½ a cup of water,
- ¼ a cup of glucose,
- ¼ a cup of butter,
- 2 ½ ozs of Baker's Premium Chocolate,
- 2 teaspoonfuls of vanilla,
- ½ a pound of Baker's "Dot" Chocolate.

Put the sugar, water, glucose and butter over the fire; stir until the sugar is melted, then cook to the soft ball degree, or 236° F.; pour on a damp marble and leave until cold; then pour on the Premium Chocolate, melted over hot water, and with a spatula turn to a cream. This process is longer than with the ordinary fondant. Cover the chocolate fondant with a bowl and let stand for thirty minutes; knead well and set over the fire in a double boiler; add the vanilla and stir until melted. The mixture is now ready to be dropped into small impressions in starch; when cold and brushed free of starch dip in "Dot" Chocolate. When dropping the chocolate mixture into the starch it should be just soft enough to run level on the top. If too soft it will not hold its shape in coating.

FONDANT FOR SOFT CHOCOLATE CREAMS

- 2 ½ cups of sugar,
- 1/3 a cup of glucose (pure corn syrup),
- 1 cup of water.

Put the sugar, glucose and water over the fire and stir until boiling, then wash down the sides of the saucepan, cover and finish cooking as in making ordinary fondant. Let cook to 238° F. Turn the syrup onto a damp marble or platter and *before it becomes cold* turn to a cream with a wooden spatula. When the fondant begins to stiffen, scrape at once into a bowl and cover with a damp cloth, but do not let the cloth touch the fondant. Use this fondant in the following recipes.

ROSE CHOCOLATE CREAMS

- Fondant,
- Damask rose color-paste,
- ½ to 1 whole teaspoonful of rose extract,
- ½ a pound of Baker's "Dot" Chocolate.

Put a part or the whole of the fondant into a double boiler over boiling water. With the point of a toothpick take up a little of the color-paste and add to the fondant; add the extract and stir until the mixture is hot, thin and evenly tinted. With two teaspoons drop the mixture into impressions made in starch; it should be hot and thin enough to run level on top. When the shapes are cold, remove from the starch, brush carefully and coat with "Dot" Chocolate.

PISTACHIO CHOCOLATE CREAMS

- Fondant,
- Green color-paste,
- 1 teaspoonful of vanilla extract,
- 1/8 a teaspoonful of almond extract,
- Pistachio nuts in slices and halves,
- ½ a pound of Baker's "Dot" Chocolate.

Using green color-paste, vanilla and almond extract mold the fondant in long shapes. Put a bit of nut in each impression, before filling it with fondant. When firm coat with "Dot" Chocolate and set half a pistachio nut on top.

SURPRISE CHOCOLATE CREAMS

- Fondant,
- Candied or Maraschino cherries,
- Flavoring of almond or vanilla,
- Chopped peanuts,
- ½ a pound of Baker's "Dot" Chocolate.

Melt the fondant over hot water and add the flavoring. Put a bit of cherry in the bottom of each starch impression, then turn in the melted fondant, to fill the impressions and have them level on the top. Let the chocolate, broken in bits, be melted over warm water, then add as many chopped peanuts as can be well stirred into it; let cool to about 80° F. and in it drop the creams, one at a time; as coated dispose them on table oil cloth or waxed paper.

CHOCOLATE PEANUT BRITTLE

- 1 ½ cups of sugar,
- 2/3 a cup of water,
- ½ a cup of glucose (pure corn syrup),
- 2 level tablespoonfuls of butter,
- ½ a pound of *raw* shelled peanuts,
- 1 teaspoonful of vanilla extract,
- 1 level teaspoonful of soda,
- 1 tablespoonful of cold water,
- ½ a pound or more of Baker's "Dot" Chocolate.

Put the sugar, water and glucose over the fire; stir till the sugar is dissolved; wash down the sides of the saucepan with a cloth or the fingers dipped in cold water, cover and let boil three or four minutes, then uncover and let cook to 275° F. (when a little is cooled and chewed it clings but does not stick to the teeth) add the butter and peanuts and *stir constantly* until the peanuts are nicely browned (or are of the color of well roasted peanuts). Dissolve the soda in the cold water, add the vanilla and the soda and stir vigorously. When the candy is through foaming, turn it onto a warm and well-oiled marble or platter. As soon as it has cooled a little on the edges, take hold of it at the edge and pull out as thin as possible. Loosen it from the receptacle at the center by running a spatula under it, then turn the whole sheet upside down, and again pull as thin as possible. Break into small pieces and when cold coat with "Dot" Chocolate prepared as in previous recipes. Half of a roasted peanut may be set upon each piece as coated. Note that the peanuts used in the brittle are raw. The small Spanish peanuts are the best for this purpose. After the peanuts are shelled, cover them with boiling water, let boil

up once, then skim out and push off the skin, when they are ready
to use.

316

CHOCOLATE POP CORN BALLS

- 1 ½ cups of sugar,
- 1/3 a cup of glucose,
- 2/3 a cup of water,
- 1/3 a cup of molasses,
- 3 tablespoonfuls of butter,
- 3 squares of Baker's Premium Chocolate,
- 1 teaspoonful of vanilla extract,
- About 4 quarts of popped corn, well salted.

Set the sugar, glucose and water over the fire, stir until the sugar is melted, then wash down the sides of the saucepan, cover and let boil three or four minutes, then remove the cover and let cook without stirring to the hard ball degree; add the molasses and butter and stir constantly until brittle in cold water; remove from the fire and, as soon as the bubbling ceases, add the chocolate, melted over hot water, and the vanilla; stir, to mix the chocolate evenly through the candy, then pour onto the popped corn, mixing the two together meanwhile. With buttered hands lightly roll the mixture into small balls. Press the mixture together only just enough to hold it in shape. Discard all the hard kernels in the corn. Have the corn warm and in a warm bowl.

CHOCOLATE MOLASSES KISSES

- 2 cups of coffee A sugar,
- 1/3 a cup of glucose, (pure corn syrup),
- 2/3 a cup of water,
- 1 cup of molasses,
- 2 tablespoonfuls of butter,
- ¼ a teaspoonful of salt,
- 4 ounces of Baker's Premium Chocolate,
- 1 tablespoonful of vanilla extract, or
- 1 teaspoonful of essence of peppermint.

Put all the ingredients, save the salt, chocolate and flavoring, over the fire; let boil rapidly to 260°F., or until brittle when tested in cold water. During the last of the cooking the candy must be stirred constantly. Pour onto an oiled platter or marble; pour the chocolate, melted over hot water, above the candy; as the candy cools on the edges, with a spatula or the fingers, turn the edges towards the center; continue this until the candy is cold enough to pull; pull over a hook until cold; add the flavoring, a little at a time, during the pulling, cut in short lengths and wrap in waxed paper.

WALTER BAKER & CO., Ltd.

ESTABLISHED 1780

This House has grown to be the largest of its kind in the world and it has achieved that result by always maintaining the highest standard in the quality of its cocoa and chocolate preparations and selling them at the lowest price for which unadulterated articles of high grade can be put upon the market. Under cover of a similarity in name, trade-mark, label or wrapper, a number of unscrupulous concerns have, within recent years, made attempts to get possession of the great market won by this House, by trading on its good name—selling to unsuspecting consumers goods of distinctly inferior quality by representing them to be the products of the genuine "Baker's." The quantity of goods sold in this way is not so much of an injury to us as the discredit cast upon our manufactures by leading some consumers to believe that these fraudulent articles are of our manufacture and that we have lowered the high standard maintained for so many years. It is difficult to bring the fraud home to all consumers, as those who are making use of it seek out-of-the-way places where deception will the more easily pass.

We have letters from housekeepers who have used the genuine Baker goods for years, expressing their indignation at the attempts of unscrupulous dealers to foist upon them inferior and adulterated articles by fraudulently representing them to be of our manufacture.

Statements in the press and in the reports of the Pure Food Commissioners show that there are on the market at this time many cocoas and chocolates which have been treated with adulterants, more or less injurious to health, for the purpose of cheapening the cost and giving a fictitious appearance of richness and strength. The safest course for consumers, therefore, is to buy goods bearing the name and trade-mark of a well-known and reputable manufacturer, and to make sure by a careful examination that they are getting what they order.

> **Our Cocoa and Chocolate Preparations are ABSOLUTELY PURE—free from coloring matter, chemical solvents, or adulterants of any kind, and are therefore in full conformity to the requirements of all National and State Pure Food Laws.**

We have behind us one hundred and twenty-nine years of successful manufacture, and fifty-two highest awards from the great industrial exhibitions in Europe and America.

We ask the cooperation of all consumers who want to get what they order and what they pay for to help us—as much in their own interest as ours—in checking these frauds.

WALTER BAKER & CO., Ltd.

Our registered guarantee under National Pure Food Laws is Serial No. 90.

WALTER BAKER & Co.'s Cocoa and Chocolate Preparations

BAKER'S BREAKFAST COCOA

FAC-SIMILE OF ½ LB. CAN.

In 1-5 lb., 1-4 lb., 1-2 lb., 1 lb. and 5 lb. tins

This admirable preparation is made from selected cocoa, from which the excess of oil has been removed. It is *absolutely pure,* and it is *soluble.* It has *more than three times the strength* of cocoa mixed with starch, arrowroot or sugar, and is, therefore, far more economical, *costing less than one cent a cup.* It is delicious, nourishing, strengthening, *easily digested,* and admirably adapted for invalids as well as for persons in health.

No alkalies or other chemicals or dyes are used in its preparation.

Trade-Mark on every package

BAKER'S CHOCOLATE

FAC-SIMILE OF ½ LB. PACKAGE.

In 1-4 and 1-2 lb. cakes, 1 lb. packages, blue wrapper, yellow label

It is the pure product of carefully selected cocoa beans, to which nothing has been added and from which nothing has been taken away. Unequalled for smoothness, delicacy and natural flavor. Celebrated for more than a century as a nutritious, delicious and flesh-forming beverage. The high reputation and constantly increasing sales of this article have led to imitations on a very extensive scale. To distinguish their product from these imitations Walter Baker & Co., Ltd., have enclosed their cakes and pound packages in a new envelope or case of stiff paper, different from any other package. The color of the case is the same shade of deep blue heretofore used on the Baker packages, and no change has been made in the color (yellow) and design of the label. On the outside of the case, the name of the manufacturer is prominently printed in white letters. On the back of every package a colored lithograph of the trademark, "La Belle Chocolatière" sometimes called the Chocolate Girl, is printed. Vigorous proceedings will be taken against anyone imitating the package.

Trade-mark on every package

BAKER'S VANILLA CHOCOLATE

In 1-2 lb. and 1-6 lb. cakes and 5c and 10c packages,

is guaranteed to consist solely of choice cocoa and sugar, flavored with pure vanilla beans. Particular care is taken in its preparation, and a trial will convince one that it is really a delicious article for eating or drinking. It is the best sweet chocolate in the market. Used at receptions and evening parties in place of tea or coffee. The small cakes form the most convenient, palatable and healthful article of food that can be carried by bicyclists, tourists and students.

Trade-mark on every package

CARACAS CHOCOLATE

In 1-8 and 1-4 lb. packages

A delicious article. Good to eat and good to drink. It is one of the finest and most popular sweet chocolates on the market, and has a constantly increasing sale in all parts of the country. If you do not find it at your grocer's, we will send a quarter-pound cake by mail, prepaid, on receipt of 10 cents in stamps or money.

Trade-mark on every package

CENTURY CHOCOLATE

In 1-4 lb. packages

A fine vanilla chocolate for eating or drinking. Put up in very artistic wrappers.

Trade-mark on every package

AUTO-SWEET CHOCOLATE

In 1-6 lb. packages

A fine eating chocolate, enclosed in an attractive wrapper with an embossed representation of an automobile in colors.

Trade-mark on every package

GERMAN SWEET CHOCOLATE

FAC-SIMILE 1/4 LB. PACKAGE.

In 1-4 lb. and 1-8 lb. packages

is one of the most popular sweet chocolates sold anywhere. It is palatable, nutritious and healthful and is a great favorite with children.

Beware of imitations. The genuine is stamped: "S. German, Dorchester, Mass."

Trade-mark (La Belle Chocolatière) on every package

DOT CHOCOLATE

In 1-2 lb. cakes; 12 lb. boxes

A high grade chocolate specially prepared for home-made candies, and for sportsmen's use. If you do not find it at your grocer's write to us and we will put you in the way of getting it.

In "The Way of the Woods—A Manual for Sportsmen" Edward Breck, the author, says:

> "Chocolate is now regarded as a very high-class
> food on account of its nutritive qualities. * * * * * A
> half cake will keep a man's strength up for a day
> without any other food. I never strike off from
> camp by myself without a piece of chocolate in my
> pocket. Do not, however, have anything to do with
> the mawkishly sweet chocolates of the candy
> shops or the imported milk chocolate, which are
> not suited for the purpose. We have something
> better here in America in Walter Baker & Co.'s
> "Dot" brand, which is slightly sweetened."

CRACKED COCOA OR COCOA NIBS

In 1-2 lb. and 1 lb. packages, and in 6 lb. and 10 lb. bags

This is the freshly roasted bean cracked into small pieces. It contains no admixture, and presents the full flavor of the cocoa-bean in all its natural fragrance and purity. When properly prepared, it is one of the most economical drinks. Dr. Lankester says cocoa contains as much flesh-forming matter as beef.

Trade-mark on every package

SOLUBLE COCOA

This is a preparation for the special use of druggists and others in making hot or cold soda. It forms the basis for a delicious, refreshing, nourishing and strengthening drink.

It is perfectly soluble. It is absolutely pure. It is easily made. It possesses the full strength and natural flavor of the cocoa-bean. No chemicals are used in its preparation.

The directions for making one gallon of syrup are as follows:

- 8 ounces of soluble cocoa,
- 8 ½ pounds of white sugar,
- 2 ½ quarts of water.

Thoroughly dissolve the cocoa in hot water, then add the sugar, and heat until the mixture boils. Strain while hot. After it has become cool, sugar may be added if desired.

The Trade is supplied with 1, 4 or 10 lb. decorated canisters.

Trade-mark on every package

CHOCOLATE FOR CONFECTIONERS' USE

Liquid Chocolates — plain, sweet, light, medium and dark.

Soluble Cocoa — for hot or cold soda.

Absolutely Pure — *free from coloring matter, chemical solvents, or adulterants of any kind, and therefore in full conformity to the requirements of all National and State Pure Food Laws.*

VANILLA TABLETS

These are small pieces of chocolate, made from the finest beans, and done up in fancy foil. The packages are tied with colored ribbons, and are very attractive in form and delicious in substance. They are much used for desserts and collations, and at picnics and entertainments for young people. They are strongly recommended by physicians as a healthy and nutritious confection for children.

Trade-mark on every package

COCOA-BUTTER

In 1-2 lb. and 1-5 lb. cakes, and in metal boxes for toilet uses

One-half the weight of the cocoa-bean consists of a fat called "cocoa-butter," from its resemblance to ordinary butter. It is considered of great value as a nutritious, strengthening tonic, being preferred to cod-liver oil and other nauseous fats so often used in pulmonary complaints. As a soothing application to chapped hands and lips, and all irritated surfaces, cocoa-butter has no equal, making the skin remarkably soft and smooth. Many who have used it say they would not for any consideration be without it. It is almost a necessary article for every household.

Trade-mark on every package

COCOA-SHELLS

In 1 lb. and 1-2 lb. packages

Cocoa-shells are the thin outer covering of the beans. They have a flavor similar to but milder than cocoa. Their very low price places them within the reach of all; and as furnishing a pleasant and healthy drink, they are considered superior to tea and coffee.

Packed *only* in 1 lb. and 1/2 lb. papers, with our label and name on them.

Trade-mark on every package

CACAO DES AZTÈQUES

In boxes, 6 lbs. each; 1-2 lb. bottles

A compound formerly known as *Racabout des Arabes*; a most nutritious preparation; indispensable as an article of diet for children, convalescents, ladies, and delicate or aged persons. It is composed of the best nutritive and restoring substances, suitable for the most delicate system. It is now a *favorite breakfast beverage for ladies and young persons*, to whom it gives freshness and *embonpoint*. It has solved the problem of medicine by imparting something which is easily digestible and at the same time *free from the exciting qualities* of coffee and tea, thus making it especially desirable for nervous persons or those afflicted with weak stomachs.

It has a very agreeable flavor, is easily prepared, and has received the *commendation of eminent physicians* as being the best article known for convalescents and all persons desiring a *light, digestible, nourishing and strengthening food.*

NO OTHER FOOD PRODUCT
HAS A LIKE RECORD.
WALTER BAKER & CO.LTD
ESTABLISHED 1780.
52 HIGHEST AWARDS.